Light Pollution as a New Risk Factor for Human Breast and Prostate Cancers

Abraham Haim · Boris A. Portnov

Light Pollution as a New Risk Factor for Human Breast and Prostate Cancers

Springer

Abraham Haim
Center for Interdisciplinary Research in Chronobiology
University of Haifa
Haifa
Israel

Boris A. Portnov
Department of Natural Resources & Environmental Management
University Haifa
Haifa
Israel

ISBN 978-94-007-6219-0 ISBN 978-94-007-6220-6 (eBook)
DOI 10.1007/978-94-007-6220-6
Springer Dordrecht Heidelberg New York London

Library of Congress Control Number: 2013931277

Printed on acid-free paper

Springer is part of Springer Science+Business Media (www.springer.com)

Preface

Humans are diurnal organisms whose biological clock and temporal organization depend on natural light/dark cycles. Throughout evolution, changes in the photoperiod were a signal for seasonal acclimatization of physiological and immune systems as well as of behavioral patterns. The invention of electrical light bulbs created more opportunities for work and leisure. However, exposure to artificial light at night (LAN) affects our biological clock, and suppresses pineal melatonin (MLT) production.

Knowledge accumulated in the past decades and our better understanding of eye photoreceptors and the discovery of melanopsin in the bipolar ganglions gave us a better perspective on light intensity and light spectrum in relation to the entrainment of our biological clock and the importance of events with timing.

In many electrical light bulbs used today and considered "environmentally friendly," electrical energy is converted into short wavelength illumination thus increasing the light intensity to the levels we have not been used to in the past. Such illumination effectively becomes "light pollution" which disrupts pineal melatonin (MLT) production. Among its other properties, MLT is an antioncogenic agent, and therefore, its suppression increases the risks of developing breast and prostate cancers (BC&PC).

To the best of our knowledge, this book is the first authored book which attempts to address the linkage between light pollution and BC&PC in humans. It explains several state-of-the-art theories, linking light pollution with BC&PC. It also illustrates research hypotheses about health effects of light pollution using the results of animal models and population-based studies.

Abraham Haim
Boris A. Portnov

Contents

Chapter 1
Introduction

Some people will never learn anything, for this reason, because they understand everything too soon.

Alexander Pope

The levels of health care and living conditions in many regions of the world have improved considerably in the past decades, especially in urban areas. Yet, modern urbanized and industrialized ecosystems are not necessarily the healthiest places to live in. In addition to many "traditional" health hazards, such as air pollution, and general stress associated with living in urban areas, yet, many new health hazards constantly emerge or are being recognized as such.

Relatively recently, for instance, exposure to low frequency electromagnetic radiation, such as radio and microwave frequencies (RF/MF) was recognized as a health risk to humans, along with exposure to various chemicals often found in urban areas, such as benzenes, detergents, endocrine disrupting chemicals (EDCs), heavy metals, and many others that are found in soils, drinking water and building materials of which our homes are built. Even more recently, shift-working, which is quite common in urban areas, has been added to the list of risk factors potentially carcinogenic to humans (ACS 2007).

In this book, which is, to the best of our knowledge, the very first authored book published on the topic, we discuss yet another potential risk factor for human breast and prostate cancers (BC&PC)—Light-at-Night (LAN), which can be termed "light pollution" or even "light toxicity."

One may ask: *How can light become a risk factor? What can be more natural to humans, as diurnal organisms, than light?*

These questions are intuitively correct, indeed. Therefore, one clarification is required: It is *not* regular daytime sunlight, to which humans have been accustomed over the years of human evolution, we are talking about. The matter is that the light we are exposed today in our homes, work places and in public spaces often differs from regular sunlight by two important properties—*timing* and *wavelength.*

Let us elaborate. Throughout the years of evolution, our human ancestors, as other mammals, were diurnal. They (normally) were active during daytime and rested at night under (normally) dark conditions. Bearing in mind that the human evolution, for a long period of time, took place close to the Equator, our ancestors followed close to 12 h of light–12 h of darkness (12L–12D) cycles.

A. Haim and B. A. Portnov, *Light Pollution as a New Risk Factor for Human Breast and Prostate Cancers*, DOI: 10.1007/978-94-007-6220-6_1,

Although humans have always been attempting to prolong the light part of the day, by whatever means locally available—burning wood, animal fat, organic and mineral oils, etc. (especially after they moved from the equatorial areas to places with short days and long nights), —possibilities for nighttime activities under such limited lighting sources, were rather limited.

The situation has changed dramatically in the past 120–130 years with the invention of an electrical light bulb, demonstrated to the public nearly simultaneously by Joseph Swan in the UK and Thomas Alva Edison in the USA in 1879.

Since then, electrical light bulbs and electricity, as an energy source for illumination, have become more reliable and affordable. Following these technological developments, electric lights proliferated widely across the globe, reaching even the most remote peripheral regions and rural areas of developing countries.

As a result of this proliferation, which has started, as we should emphasize, only 120–130 years ago, humans across the globe are no longer "tied" today to the traditional 12L–12D cycles, but can be active around the clock, if they chose so. Supported by artificial illumination, we can be active at night and rest during the day, quite contrary to our diurnal nature, "programmed" by the years of evolution.

In addition to these *changes in the temporal activity patterns*, which artificial illumination enables, our eyes are often exposed to *high light intensity when they are supposed to have been exposed to very low light intensity* (if at all), that is, after sunset and even at night.

LAN often penetrates our bedrooms from outdoor sources through fenestrations in our walls. Outdoor LAN sources include streetlights, as well as lights from billboards, stadiums, shopping centers and other brightly illuminated public buildings and monuments, neighboring buildings, moving vehicles, etc. (Fig. 1.1).

In addition, LAN is often present in our bedrooms when we sleep. It comes from nightlights, working TV sets, computers and other indoor equipment and devices we do not bother (or do not want) to turn off. Moreover, light indicators, including digital clocks, are often switched on, as well as "standby" lights on other electronic devices, such as computers, cellular phones, air conditioners, routers, TVs, DVD players, etc.

In addition, we are also exposed to LAN in our workplaces (especially people who work night shifts), as well as in places of nighttime entertainment—sport and cultural facilities, movie theaters, etc. The *timing of our light exposure* today is thus quite different from what we were evolutionary "programmed."

An additional difference between natural daylight and artificial lights that we commonly use today should also be mentioned. This is the *difference in the wavelength* emitted by these light sources from that of natural sunlight.

The matter is that visible *sunlight* is characterized by a daily changing wavelength in a wide range. In contrast, many artificial light sources, we use today, emit short wavelengths of visible light with a constant predominant wavelength between 450 and 500 nm.

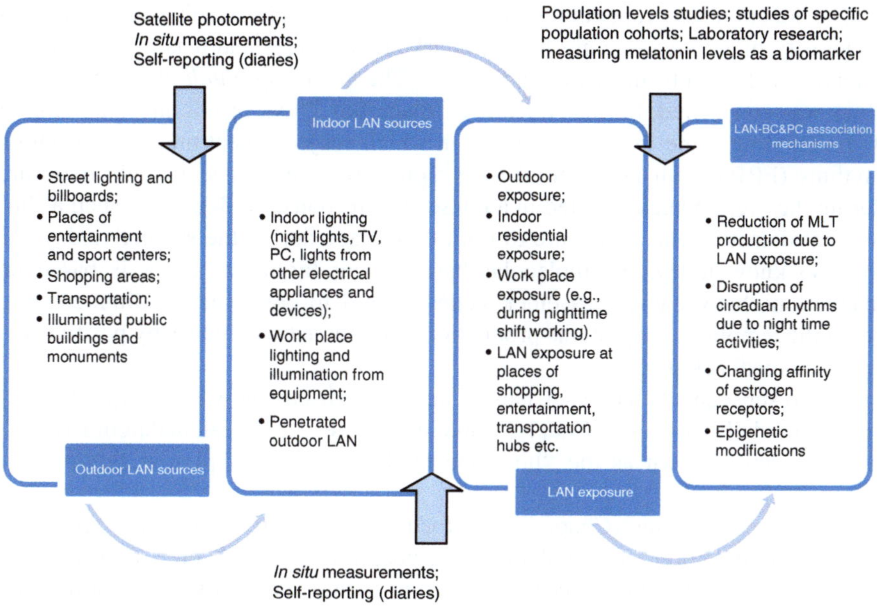

Fig. 1.1 Links between LAN and breast-prostate cancers (BC&PC) —research hypothesis (after Haim and Portnov 2011)

> Do these nightlight-enabled alterations in our life styles and newly introduced light sources interfere with our "preprogrammed" daily cycles? Can these changes and newly introduced LAN sources potentially suppress MLT production, weaken our immune system, and thus expose us to additional health risks, such as BC&PC?

In the rest of this book we shall attempt to answer these questions.

We start first with a brief discussion of visual light and its physicochemical properties. We shall also review selected characteristics of light bulbs used today for indoor and outdoor lighting and spectra of different light sources. These concepts and notions will be discussed in the next chapter (Chap. 2 written by Dr. Fabio Falchi).

In Chap. 3, we discuss the light and dark cycles (photoperiodicity) on Earth as the underlying basis of our temporal organization. As we shall emphasize, most living terrestrial organisms anticipate the time of activity due to the existence of the endogenous biological clock entrained to exogenous environmental photic changes. As we shall argue, our daily rhythms can be described as an orchestra in which the harmony of all instruments is maintained by the single conductor, with the biological clock (the master oscillator) carrying out this function. The biological clock also works as a calendar helping living organisms to anticipate seasonal changes and to acclimatize their physiological and immune systems, as well as their behavioral patterns to the approaching season. For its temporal organization, the biological clock uses the signals of light intensity and wave length detected by the retina in our eyes, but not only as in the case of vitamin D when the light signal is picked by the skin.

The location and functioning of the biological clock, which is entrained, as previously mentioned, mainly by light/dark (L/D) cycles and acts as our internal calendar, will then be discussed in Chap. 4. The *modus operandi* of our biological clock was considered, for many years, a "black box." Our recent understanding of its functioning came from experiments in which subjects, kept under free running rhythms (FRR), with no external time signal, were entrained to light exposure during the night, which resulted in a phase shift response (PSR), while such a shift was not detected when exposure to light took place during the subjective day. As we also know today, the molecular basis of our biological clock is based on interactions between various clock genes, paired to create hetero-dimmers, and being the templates for proteins production that interact with genes by positive and negative feedback loops.

In two subsequent chapters, Chaps. 5 and 6, we discuss how LAN exposure may affect our daily rhythms (Chap. 5) and what role the pineal melatonin (MLT) hormone, also known as the "hormone of darkness," may play in this process (Chap. 6).

In the next chapter (Chap. 7) we focus on the introduction and spread of artificial illumination. As we shall argue, people knew about electricity for centuries. However, the first successful attempt to use electricity for lighting took place only about two centuries ago and is credited to Sir Humphrey Davy, who discovered in 1801 the incandescence of an energized conductor. Yet, the idea of using electricity for lighting "took off" only after the American inventor Thomas Alva Edison developed his deep vacuum incandescent light bulb with a carbon cotton filament. Since then, both light bulbs and electricity production have become relatively cheap and more reliable. As a result of rapid electricity proliferation, electric lighting has substituted traditional lighting sources, making human populations virtually independent of natural L/D cycles.

In the next three chapters (Chaps. 8–10) we deal, in brief, with the biological definition of light pollution (Chap. 8), the role of light pollution as a general stressor (Chap. 9) and the effects of light pollution on animal rhythms and ecology (Chap. 10). As we shall emphasize, different theories can be used to explain the association between LAN exposure and BC&PC. The direct effect of LAN on the eye retina and resulting infractions of human daily rhythms by disrupting pineal MLT production and secretion, is only one of them. LAN can also act as a general stressor. LAN, through MLT suppression, can also become a cause of changing affinity of estrogen receptors or modify DNA global levels of methylation thus causing DNA changes epigenetically.

Empirical evidence accumulated to date about the links between LAN and BC&PC is discussed and summarized in Chap. 11. As we shall note, a possibility that human body may be affected by ambient light was raised, apparently for the first time, by the Israeli physician Philip Cohen in 1970. As we know today (from clinical experiments, case-control and population levels studies), the link between exposure to artificial light and its potential health risks may be attributed to two interdependent mechanisms—inhibition of MLT secretion from the pineal gland by direct exposure of human vision system to LAN and disruption of daily rhythms by nighttime activities. Empirical evidence for these effects is discussed in Chap. 11, supported by reviewing numerous empirical studies carried out to date.

Worldwide patterns of BC&PC are discussed in Chap. 12. As we shall demonstrate, geographic patterns of these cancers worldwide are surprisingly similar, with higher rates of both BC&PC observed in developed countries and lower rates elsewhere. The incidence rates of these cancers are also higher in more extreme latitudes, suggesting a possibility that such rates are related, among other factors, to LAN, considering that in extreme latitudes artificial light is often used to compensate for a shortage of natural illumination. Our analysis supports such a possibility, especially for PC.

Light pollution and its associations with BC&PC in population-level studies are discussed in Chap. 13, using two specific case studies, which attempted to link digital maps of nighttime illumination captured by satellite sensors with BC&PC incidence rates, thus helping to demonstrate the association potentially existing between them. As we acknowledge, nighttime satellite imagery has been used before, for mapping sky brightness and built surfaces, construction of "global poverty" maps, estimation of ecological footprints of different countries and country specific electrification rates, spectral identification of lights, monitoring forest fires, and for many other development tasks. However, the idea to link the digital satellite images with place-specific incidence rates of BC&PC was originated in two studies spearheaded by the authors of this book.

As we assumed from the outset of the analysis, urban populations residing in highly illuminated areas (such as e.g., central London, Paris, Tel Aviv or NYC) are exposed to LAN not only in their bedrooms but also in many other places, and from a variety of other light sources, which residents of smaller towns and of rural areas do not have, at least, on such a scale. The daily rhythms of the residents of such major populations may also be disrupted by various nighttime activities, such as leisure and entertainment, and employment in businesses working after dusk. In this sense, satellite photometry helps to capture these additional LAN-associated risks. Our underlying research hypothesis was relatively simple: *If there is a significant relationship between population LAN exposure and BC&PC incidence rates, then there should be a significantly strong association between LAN intensities and BC&PC, but not with other cancers, such as colon, larynx, lung, etc.*

Population-based and individual-level studies have their own advantages and disadvantages. Global studies of large populations may provide a high degree of generality and thus help to capture the effects of low exposures by comparing a wide range of differently exposed subjects. However, population studies are generally weak in supporting causality. They also often overlook detailed characteristics, such as hereditary factors, residential history and occupational risks. Other biases may also affect the results of population-level studies of the LAN-BC&PC association, including ecological fallacy, recall bias, and the eyelid effect. These potential bias sources and ways of mitigating them are discussed, in some detail, in Chap. 14.

As we further argue, although individual-level studies may provide rich details about studied subjects, there are obvious limits to the degree of generality and statistical power that a semi-random sample of a few dozens of subjects, for a short period of time (several days), at a certain time in the year, lacking a matched control, can provide. In this setting, population level studies can assist in initial hypothesis testing, assuming that the results of such studies can be followed up by higher resolution studies in humans and by using animal models to understand mechanisms down to the cellular level.

Although most of the studies reviewed in this book relate to BC&PC, there is a rapidly increasing body of evidence, coming from both laboratory research and epidemiological studies, that LAN exposure may also be linked to other health effects, such as e.g., hypertension, obesity, sleep disruption and mental disorders. The mechanism of these potential associations may be similar to those we discuss referring to the LAN-BC&PC links, namely MLT suppression and daily rhythms disruption. Therefore, in the concluding chapter (Chap. 15), we review, in brief, scientific evidence which has become recently available on other, non-BC&PC related, effects of LAN exposure.

As we also point out in Chap. 15, uncontrolled and rapidly increasing exposure to LAN may present a new and serious health challenge for ever increasing human population worldwide. However, the main message we attempt to deliver in this book is not that humans should go back to the "pre-Edison" era of "nighttime darkness." Such calls would be both counter-productive and unrealistic. Instead we should think of adhering, whenever possible, to our "evolutionary-preprogrammed" diurnal life cycles. We should also try to implement sustainable illumination policies, both in our homes and in public places. In particular, we should refrain from using short wavelength light sources, aggressively brought into our private and public domains in the name of "energy saving." Although such illumination sources can save energy, their adverse health effects, in the long run, can greatly outweigh, in our view, any energy saving benefits such light sources can potentially bring today and tomorrow.

Finally we would like to note that not all the aspects of the LAN-BC&PC association are yet clear, and answers to many questions are still pending. Therefore, research on this important topic should, undoubtedly, continue, using various methodologies and comprehensively designed studies.

Part I
Artificial Light and Human Temporal Organization

Chapter 2
Artificial Light and its Physicochemical Properties

Abstract Light is a small part of the electromagnetic spectrum, from violet to red, to which our eyes are responsive. Photometry measures light using several units, including the candela for intensity, the lumen for flux, the lux for illuminance and the candela per square metre for luminance. Vision at relatively high illumination levels is called photopic, when our eyes mainly use cones; in the dark, vision is called scotopic. Recently, a non-visual photoreceptor with peak sensitivity in the blue part of the spectrum has been discovered which regulates our circadian rhythms.

Keywords Light spectrum · Spectral range · Electromagnetic radiation · Photometry · Light flux · Light units · Light intensity · Photopic and scotopic vision · Bulbs · Meltopic efficacy · Photoreceptors · Illuminance · Artificial lighting · Eye sensitivity

> *Philosophy is written in that great book which ever lies before our eyes—I mean the universe—but we cannot understand it if we do not first learn the language and grasp the symbols, in which it is written.*
>
> Galileo Galilei (The Assayer, 1623)

Visible Light

The portion of electromagnetic radiation visible to the human eye we call light. Light can be described as an electromagnetic wave, like radio waves, microwaves, infrared radiation, ultraviolet radiation, X-rays and gamma-rays. All these

---

'La filosofia è scritta in questo grandissimo libro che continuamente ci sta aperto innanzi a gli occhi (io dico l'universo)...' Galileo Galilei, 'Il Saggiatore'.
The chapter is contributed by Dr. Fabio Falchi (ISTIL-Light Pollution Science and Technology Institute, Thiene, Italy).

A. Haim and B. A. Portnov, *Light Pollution as a New Risk Factor for Human Breast and Prostate Cancers*, DOI: 10.1007/978-94-007-6220-6_2,

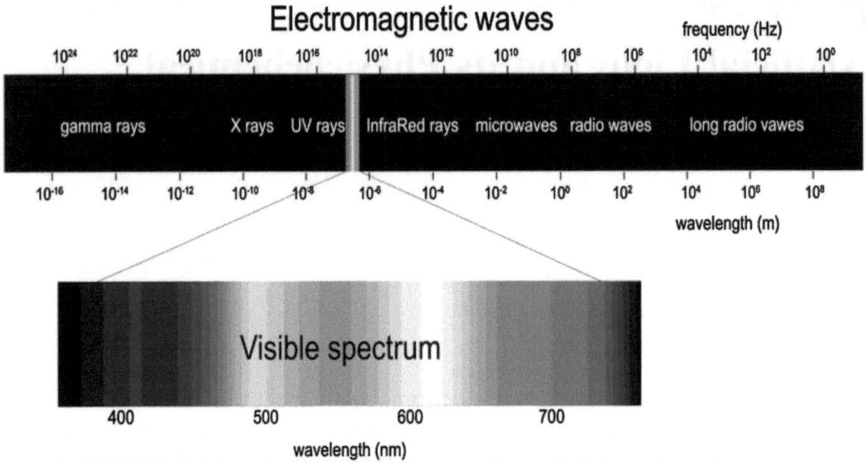

Fig. 2.1 The visible spectrum as a part of the electromagnetic radiation. *Note* Our eyes are blind below about 390 nm and above about 780 nm

radiations differ by their wavelength, i.e. the distance between two subsequent crests. The borders between different rays (e.g. between X-rays and gamma rays) are gradual, with one fading into the other (Fig. 2.1).

White light, such as the light arriving from the Sun, can be separated into a spectrum by a glass prism or a diffraction grating. The spectrum of the Sun or an incandescent bulb presents a continuum transition from the shortest wavelength colour visible to the eye, which we see as violet, to the longest wavelength one, which we see as red (Table 2.1).

Our eyes are generally sensitive to wavelengths between 390–780 nm. One nanometre (symbol nm) is one billionth of a metre. The centre of the visible portion of the e-m spectrum has wavelengths about half of one-thousandth of a millimetre.

The speed of light is constant in vacuum. Its value is $c = 299792458$ m/s, i.e., nearly 300,000 km/s. The speed of light is lower in all the media, such as water and glass. The speed of light c, its wavelength λ and its frequency v are related by this simple equation:

$$c = \lambda v \tag{2.1}$$

Table 2.1 Approximate spectral range of pure colours with one fading into the next with continuity (after Benenson et al. 2002)

Colour	Frequency (10^{12} Hz)	Wavelength (nm)
Violet	659–769	390–455
Blue	610–659	450–492
Green	520–610	492–577
Yellow	503–520	577–597
Orange	482–503	597–622
Red	384–482	622–780

Light, being electromagnetic radiation, represents both a wave and particle nature. Some phenomena of light can be explained by its wave aspects (e.g. diffraction and interference). Other phenomena require the use of the corpuscular nature of light (e.g. photoelectric effect, Compton effect, etc.).

The particle of light is called photon. Each photon carries an energy that depends on its frequency:

$$E = h\nu \tag{2.2}$$

where h is Plank's constant, $h = 6.626069\ 57 \times 10^{-34}$ J·s (Mohr et al. 2011).

Photometry: The Science of Measuring Light

There are different ways to measure light, using different units of measure. It is possible to use the unit of energy (joule) and power (watt), but at equal power, two light fluxes of different wavelength produce two different perceptions in human eyes. For this reason we introduce the photometric units, such as the *lumen* and the *candela* that take into account the sensitivity of our eyes to the different portions of the light spectrum.

When there is sufficient light, i.e. the luminance (see below) is more than about 3 cd/m^2, our eyes use mainly the cone photoreceptors in what is called *photopic* vision (see Fig. 2.2).

Light Flux

Human peak sensitivity is at 555 nm, where our eyes see a green colour. At this wavelength one watt of radiant power produces 683 lumens of luminous flux. The

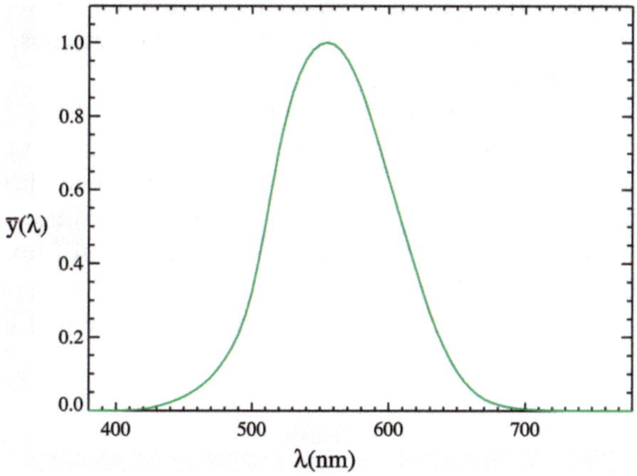

Fig. 2.2 The diurnal or *photopic* sensitivity of the human eye (sight mediated by *cones*), with the maximum normalized to 1 (Public domain graph)

lumen (symbol: lm) is the unit of measure of the light flux and it corresponds to a power, i.e. an energy over time.

One watt of light at 610 nm produces a lower, orange stimulus in our eye, about half of green light, so we get only about 340 lumen. One watt of blue light at 470 nm produces even less stimulus, about one tenth of the peak sensitivity, so it corresponds to about 70 lumen. One watt of UV radiation at 350 nm, or IR radiation at 1,000 nm, both outside the visual range, where our eye is blind, does not produce any stimulus, so one watt of UV or IR radiation corresponds to a light flux of zero lumens.

To compute the flux of a light source, given its radiant spectrum $\Phi_{e,\lambda}$ the following is used:

$$\Phi_{\text{Photopic}} = K_m \int_0^\infty \Phi_{e,\lambda} V(\lambda) d\lambda \tag{2.3}$$

where $V(\lambda)$ is the *photopic* response (Fig. 2.2) and $K_m = 683$ lm/W is the standard lumen per watt conversion factor.

Light Intensity

In the *Système Internationale d'Unités* (SI) the lumen is a derived unit. The fundamental unit is the *candela* (symbol: cd) that measures luminous intensity, equal to the luminous flux emitted by a light source in a solid angle element. The definition of candela (Bureau International des Poids et Mesures 2007) is:

> The candela is the luminous intensity, in a given direction, of a source that emits monochromatic radiation of frequency 540×10^{12} Hz and that has a radiant intensity in that direction of 1/683 watt per steradian.

The frequency of 540×10^{12} Hz corresponds to a wavelength of 555 nm, as given by Eq. 2.1. The *steradian* is the unit for the solid angle. A whole sphere seen from inside occupies a solid angle of 4π steradians. The product of intensity with solid angle gives light flux, so that:

lumen = candela·steradian

For a source with constant intensity of 1 cd in every direction (toward all the 4π solid angle, i.e. an isotropic source), the total flux emitted is $1 \text{ cd} \cdot 4\pi \cong 12.57$ lumens. Conversely, given the total flux emitted by an isotropic source, it is possible to get the light intensity by dividing the flux by 4π. For example, a frosted light bulb is almost isotropic, except toward the connection. A 100 W incandescent emits a flux of 1,750 lm, so its intensity will be about $1750/4\pi \cong 139$ cd.

Table 2.2 Approximate typical illuminances in various settings

Situation	Illuminance (lx)
Surface lighted by the Sun	100,000
Clouded daylight	1,000–10,000
Office interiors	100–1,000
Home interiors artificially lighted	30–300
Street lighting	5–50
Full moon illumination	0.2
Quarter moon illumination	0.02
Illumination by clear starry sky	0.0001
Illumination by overcast moonless night sky	0.00005

Illuminance

Illuminance is the ratio between the light flux that hits a surface and the surface area. Its unit is the *lux* (symbol lx) and it is equal to 1 lm/m^2. Examples of typical illuminances are given in Table 2.2. The illuminance produced by the Sun on a clear day is about one thousand times the light in artificially lighted home interiors, and 10 thousands more than the illuminance in lighted streets. However, this comparison is not very fair, due to the fact that we compare light in daytime to nighttime. During the night, natural lighting reaches a peak of about 0.2 lux in the few days around the full moon. Artificial lighting in streets and in home interiors is 100–1,000 times greater. During most of the lunar month, natural light is much less, so that artificial light is 100,000 to over than a million times more intense than the natural light.

Luminance

Luminance is the quantity most similar to how 'bright' we perceive a surface. If we look at two foils, one of black paper and the other of white paper, we perceive the white one as much brighter than the black, even if they are lighted in the same way, i.e. if they receive the same illuminance from a source such as the Sun. The unit for luminance is the candela divided by the square metre of the projected area (cd/m^2) (Tables 2.3, 2.4).

Efficacy

Light bulbs produce light using electric power, in most cases. The efficacy in producing light is given by the lumen/watt ratio. In theory, the highest possible efficacy for a bulb emitting a monochromatic light at 555 nm would be 683 lm/W. In practice bulbs emits light at different colours, where the eye sensitivity is lower

Table 2.3 Approximate typical luminance of some light sources (data adapted, in part, from Rea 2000)

Surface	Luminance (cd/m^2)
Sun (high in the sky in a very clear day)	1.6×10^9
Tungsten-halogen bulb	$2–4 \times 10^7$
Compact fluorescent	$1–10 \times 10^4$
60 W frosted incandescent bulb	1×10^5
Clear sky	8×10^3
Daytime landscape	$5–10 \times 10^3$
TV monitor (displaying white)	50–300
Artificially illuminated road	0.5–3

Table 2.4 Correspondence between different type of physical quantities

Quantity	Energetic or radiometric	Photonic	Photometric
Flux	watt	Photons s^{-1}	Lumen (= cd sr)
Intensity (flux in a solid angle)	$W\ sr^{-1}$	Photons $s^{-1}\ sr^{-1}$	Candela
Illuminance	$W\ m^{-2}$	Photons $s^{-1}\ m^{-2}$	Lux (= $lm\ m^{-2}$)
Luminance, (~ brightness)	$W\ m^{-2}\ sr^{-1}$	Photons $s^{-1}\ m^{-2}\ sr^{-1}$	$cd\ m^{-2}$ (= $lm\ m^{-2}\ sr^{-1}$)

Table 2.5 Selected characteristics of bulbs used for indoor and outdoor artificial lighting

Bulb	Power (W)	Flux (lm)	Efficacy (lm/W)
Incandescent	100	1,400	14
Fluorescent	21	1,900	90
Compact fluorescent	18	1,200	67
High pressure sodium	70	6,500	93
High pressure sodium	150	15,000	100
Low pressure sodium	35	4,500	130
Low pressure sodium	180	32,000	178
Warm white LED[a]	12	806	67

[a] Philips MASTER LED bulb D 12-60 W E27 2700 K 230 V

than the maximum. This, combined with other factors, such as heat production and consequent energy loss, lowers the bulb's efficacy (Table 2.5).

Are all Lumens Equal for Humans?

We can obtain flux of one lumen by using different amounts of colours, by using different bulb spectra (Fig. 2.3). The three spectra shown in Fig. 2.4, weighted by the photopic sensitivity response, give the same flux, even if the LED light has a

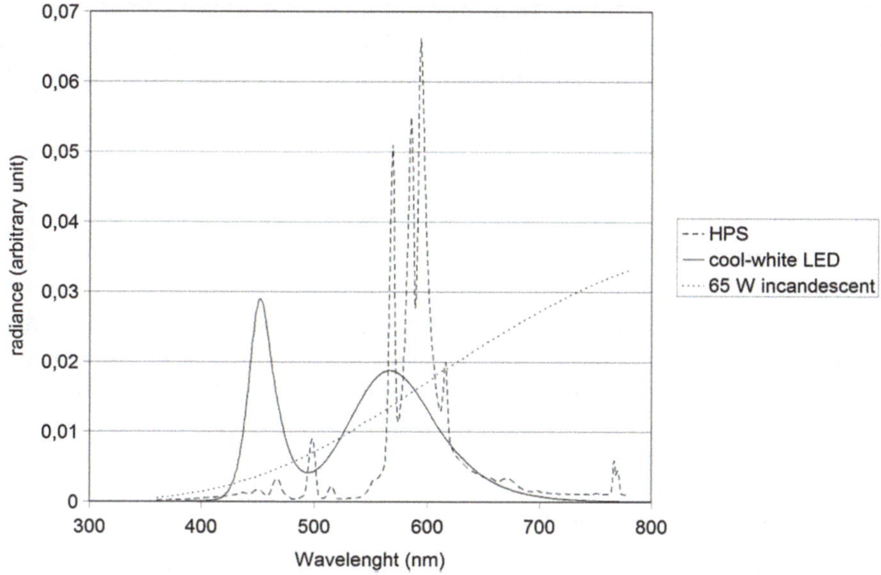

Fig. 2.3 Spectra of white LED (*solid line*), High Pressure Sodium bulb (*dashed line*) and incandescent bulb with an equal flux (Adapted from Falchi et al. 2011)

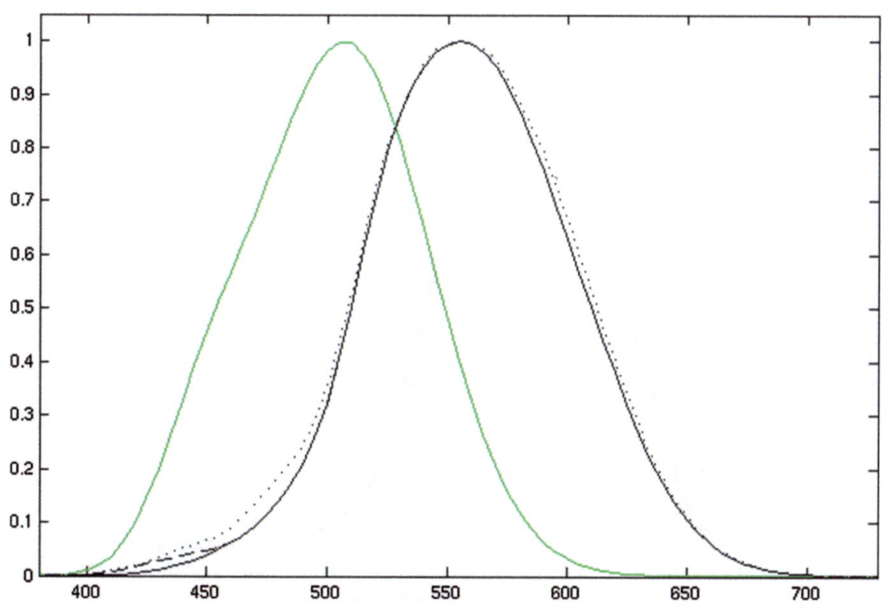

Fig. 2.4 Photopic (*black*) and scotopic (*green*) luminosity functions. The photopic includes the CIE 1931 standard (*solid*) and successive sensitivity curves by Judd-Vos (Vos 1978) (*dashed*), and by the Sharpe, Stockman, Jagla and Jägle 2005 (*dotted*). The horizontal axis is wavelength in nm (*Source* Wikimedia commons)

higher blue content, while the High Pressure Sodium bulb has a higher yellow content. Note the smooth spectrum with low blue content of an incandescent bulb that our eyes perceive as a comfortable warm white. Does our vision and physiology respond in the same way to exposure to one lux given by a red bulb, by a yellow or by a blue one?

The Scotopic Vision

Our eyes, when dark adapted, have a different sensitivity curve, more shifted toward the shorter wavelength than the photopic curve used in the photometric units. When ambient luminance is lower than about 0.01 cd/m^2 (about the luminance produced by the illumination given by full Moon) our eyes are not anymore able to distinguish colours, so we see in black and white, as our high sensitivity photoreceptors, the rods, are monochromatic, i.e. colour blind. The peak sensitivity of the dark adapted eye is at about 507 nm (Fig. 2.4).

The scotopic lumen can be introduced in analogy with Eq. 2.3:

$$\Phi_{\text{Scotopic}} = K'_m \int_0^\infty \Phi_{e,\lambda} V'(\lambda) d\lambda \tag{2.4}$$

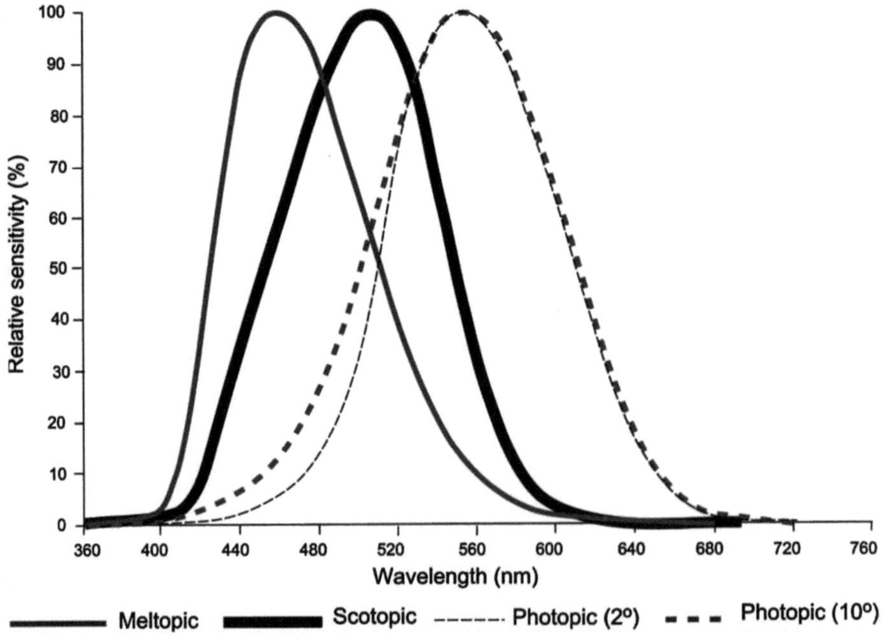

Fig. 2.5 The meltopic, scotopic and photopic sensitivity functions (Graph by David Keith after Hollan 2004)

Table 2.6 Scotopic flux to photopic flux ratio for some types of sources (after Falchi et al. 2011)

Source	S/P ratio
Low pressure sodium	0.20
High pressure sodium 70 W	0.55
Average high presure sodium	0.66
Mercury vapour 80 W	1.18
CIE illuminant A	1.41
QTH 3,100 K	1.56
Average metal halide	1.60
Flat spectrum	1.86
LED 'natural white'	3.5

where $V'(\lambda)$ is the scotopic response (Fig. 2.5) and $K'_m = 1{,}699$ lm/W is the standard lumens per watt conversion factor for scotopic response.

To take into account for the difference in the photopic and scotopic responses to flux, it can be introduced by the scotopic to photopic ratio, *S/P*:

$$S/P = \frac{\Phi_{\text{Scotopic}}}{\Phi_{\text{Photopic}}}$$

Bulbs can have the same photopic flux, but very different S/P ratios (Table 2.6).

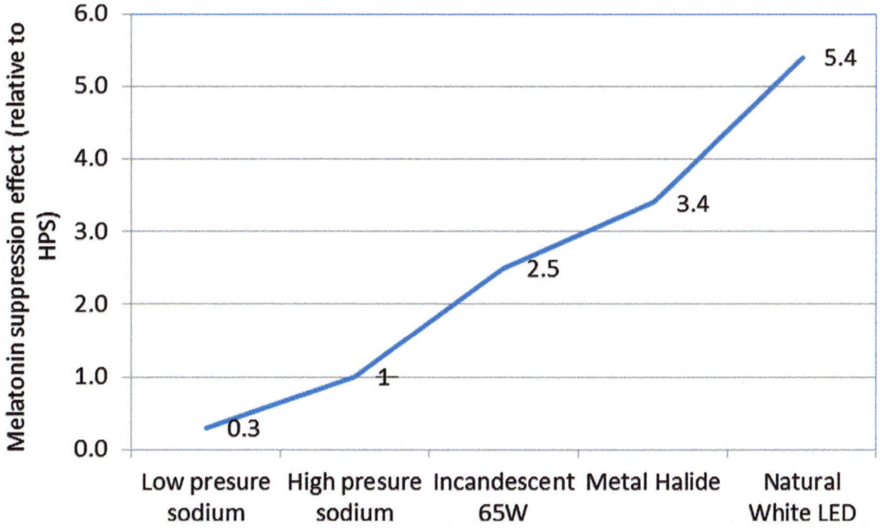

Fig. 2.6 Calculated meltopic efficacy of common types of bulbs, compared to the standard High Pressure Sodium, at equal photopic output (after Falchi et al. 2011)

'Meltopic' Sensitivity

The recently discovered Non-Image-Forming Photoreceptors (NIFPs) and the photopigment melanopsin allows our body to regulate our circadian rhythms via the exposure to light. The sensitivity curve for this type of exposure has a peak in the blue part of the e-m spectrum, around 460–480 nm (Fig. 2.5). This difference in sensitivity implies that the emission spectrum of a bulb can make a big difference in circadian disruption, even if the measured light quantities expressed in photometric units (e.g. lumen, lux) are the same. In the graph in Fig. 2.6, it is shown that a cool-white LED suppresses melatonin production more than five times compared to standard High Pressure Sodium bulb, and about 20 times more than orange Low Pressure Sodium bulb. For these differences, it is fundamental that the studies on the effects of light on life take into account the spectra of the bulbs used. It will not suffice to measure the illuminance or the flux in standard photopic lux and lumen. At the very least, the S/P ratio of the bulb used should be given.

Chapter 3
Light and Dark Cycles as a Basis of Temporal Organization

Abstract Different living organisms are active at different times of the 24-h cycle. Results of controlled experiments revealed that living organisms anticipate the time of activity due to the existence of an endogenous biological clock entrained by the exogenous environmental *Zeitgeber*. Our daily rhythms can thus be described as an orchestra in which the harmony of all instruments is maintained by the single conductor, with our biological clock carrying out such a function. The biological clock also works as a calendar helping living organisms to anticipate seasonal changes and to acclimatize their physiological systems and behavioral patterns to the approaching season. For its temporal organization, the biological clock uses the signals of light intensity detected by eyes, as well as signals of different wavelength.

Keywords 24-h cycle · Biological clock · *Zeitgeber* · Daily rhythms · Synchronization · Biological calendar · Seasonal changes · Light intensity · Light–dark Cycles · Light signals · Photoperiod

There is a time for everything, and a season for every activity under heaven.

Ecclesiastes 3:1

Time for Everything

Timing of events, both taking place on the cellular level and involving interactions between organisms and their environment, can be looked upon as a result of natural selection. Humans, as diurnal organisms, are adapted to gradual changes in light intensity between day and night and *vice versa*.

The awareness about seasonal changes finds its expression in the Old Testament:

A. Haim and B. A. Portnov, *Light Pollution as a New Risk Factor for Human Breast and Prostate Cancers*, DOI: 10.1007/978-94-007-6220-6_3,

Yea, the stork in the heaven knoweth her appointed times; and the turtle and the crane and the swallow observe the time of their coming (Jeremiah 8:7).

What is the main underlying cause of cyclic changes we experience on the daily basis? Do these cycles have anything to do with our biological clock and our internal calendar that follows annual changes known as seasons?

Ecology is the discipline that attempts to understand and explain the interrelations between living organisms and their environment (both *abiotic* and *biotic*). In doing so, ecology deals with spatial and temporal variables. However, since the beginning of the twentieth century, research attention has been mainly focused on spatial variables. Attempts to understand the temporal variables have started much later, although the researchers had been aware of the fact that organisms can be diurnal, crepuscular or nocturnal, as a mean of reducing competition. The research community was also aware of seasonality in organisms such as migration of birds and other species.

However, the ways by which environmental signals are picked up by the organism have attracted research attentions only in the past fifty years with the most significant contribution being made by *chronobiologists*, that is, scientists, studying biological rhythms and seasonality.

Discovery of the Biological Clock

In fact, the existence of the biological clock was first shown experimentally nearly 300 years ago by the French scientist Jean-Jacques d'Ortous De Mairan, who described, in 1720, the movement of the mimosa leaves. These leaves open in the morning and close in the afternoon in anticipating of the night. He also showed experimentally that these movements went on even under complete and continuous darkness, and that all leaves of the plant moved simultaneously. This was the first experimental result showing the existence of an *endogenous* biological clock in the plants.

De Mairan also noted that the rhythm of opening and closing of the mimosa leaves did not follow exactly a 24-h cycle. Based on this observation, he coined the term "circadian" which means "about a day." Already then De Mairan demonstrated the variability in the clock period among individual plants, albeit not naming it yet the "biological clock."

The study of the biological clock in animals came much later. These studies were pioneered by Jurgen Aschoff from Germany, Franz Halberg and Colin Pittendrigh from the USA. However, it took a long time to accept the new paradigm, according to which the organism can anticipate environmental changes on a daily or seasonal basis instead of simply following such changes and responding to them.

The idea of *homeostasis* was first suggested by the French physiologist Claude Bernard in the nineteenth century and defined as the ability of an animal to maintain a relatively constant internal environment. This idea revolutionized our way of understanding the relations between the organism and its environment.

The discovery of the various elements forming the circadian system in mammals and the connection between them, on the one hand, and the transfer of environmental signals entrainment, on the other, made it clear that physiological and behavioral functions tend to exhibit daily variations and that these variations, known as *daily rhythms*, are controlled by the biological clock. As demonstrated, for instance, by Konopka and Benzer (1971), such changes are based on cellular machinery controlled by genes on the cellular level, while, on the organism level, they are controlled by the endocrine system. We also face anatomical changes but they are much less common in seasonal acclimatization.

For simplicity's sake, many daily rhythms in the mammal body can be described as an orchestra in which the harmony of all instruments is maintained by the single conductor, that is, the biological clock (the central oscillator or *pacemaker*), which plays the crucial role of the coordinator. The question is: *How does our biological clock work*?

Biological Clock Functioning

The biological clock, which also acts as a calendar (Reiter 1993), is entrained by external geophysical cycles (or *Zeitgebers,* from German for "time giver," or "synchronizer"), such as light and dark (L/D) cycles, which results from the rotation of our Planet around its axis (see Fig. 3.1).

These rotations and resulting changes in daytime light intensity serve as the basis for entrainment of our biological clock, which "tunes itself up," at least twice a day, during the sunset and sunrise. Such cycles are thus the main "Zeitgeber," to which biological clocks of many terrestrial organisms entrain.

Due to the fact that the axis of our planet is not exactly perpendicular to its orbit, different ratios between the length of the light phase (*photophase*) and the dark phase (*scotophase*) in different parts of the globe are observed. These relations change by seasons of the year depending on hemisphere and latitude. The

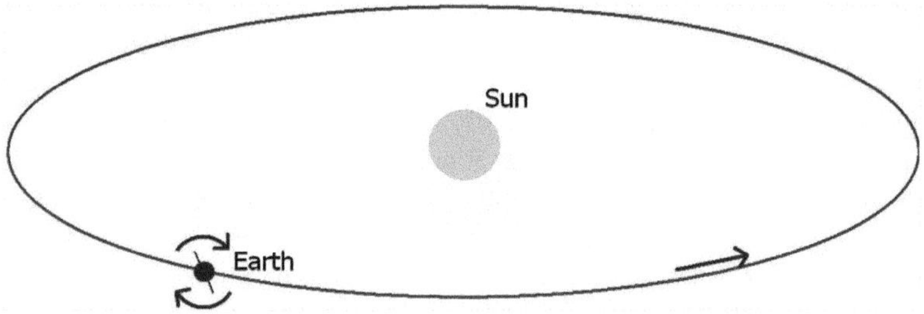

Fig. 3.1 Rotation of earth around its axis and along its orbit around the sun as a time keeper (or "Zeitgeber"), which is the underlying basis of our temporal organization

ratio between light and dark hours of the 24-h cycle is named *photoperiod* and it is used by terrestrial organisms to prepare their physiological and immune systems as well as their behavioral patterns to the coming season.

Our understanding of the mechanisms used for such changes came mainly from studies that have been carried out since the early 1960s. These studies dealt with thermoregulatory aspects, reproductive performances, anatomical changes and the functioning of the immune system (see *inter alia* Aschoff 1960, Nelson 2005).

As these studies noted, seasonal acclimatization mechanisms involved the changing levels of the pineal neuro-hormone—*Melatonin* (MLT). The fact that receptors of MLT are found on the cellular level of many organs, including white blood cells (lymphocytes), indicates its pivotal role in the normal functioning of our body including the functioning of our immune system.

An important discovery related to the response of the mammal body to photoperiod changes and the role which the eye plays in entraining the circadian system came from a study of the "blind" mole-rat *Spalax ehernbergi* who responded to an increase in the number of dark hours by increasing its thermoregulatory resistance upon exposure to low ambient temperatures (Haim et al. 1983). These results gave rise to further research that led to the better understanding of the retina structure and the existence of non-image forming photoreceptors (NIFPs), in addition to the cones and rods which are known today as image forming photoreceptors (IFPs).

Light/Dark Cycles and Human Circadian System

Responses to L/D cycles are basic events on our Planet, which we can see even in the unicellular organism *Euglena*, which has a photoreceptor enabling it to detect environmental L/D signals. Evolutionary, it may imply that our biological clock is based on a photoreceptor which detects light intensities. Through the entire evolutionary process, this basic mechanism of picking up the light and dark signals has been retained in mammals in the form of NIFPs.

The light spectrum emitted by the Sun in a certain geographical location on our Planet changes according to the time of day. The main change, we are aware of, is light intensity which is highest at noon and early afternoon. However, different wave lengths of emission also show variations during *photophase*. Thus, in the morning, noon and early afternoon hours, the dominant wave length that hits our Planet is less than 500 nm. Concurrently, in the afternoon and early evening the wave length radiation which hits our Planet is over 600 nm. Entrained to these changes, our biological clock anticipates these daily changes and is thus able to adjust the activity of various systems, by increasing or decreasing the sympathetic tone which is responsible for our alertness, body temperature and blood pressure. The secretion of MLT during the dark phase also decreases alertness and other physiological variables while increasing sleepiness.

Since light/dark cycles are a basic component of our daily and annual temporal organization, it is thus not surprising that exposure to LAN disrupts our temporal organization. The disruption of entrained daily rhythms is a phenomenon we experience during a jetlag.

By the same token, an exposure to short wave illumination in the late afternoon or early evening may inform our major oscillator that we are at midday, resulting in increasing, instead of decreasing, alertness.

In summary, for the entrainment of our daily rhythms and for the temporal organization of our biological clock, *we use signals of different light intensity and wave length detected by our eyes at the sunset and sunrise. For our seasonal organization, the most important environmental signal is the photoperiod.*

Throughout the evolution, the "calendar" entrainment of different terrestrial organisms by the photoperiod has been one of the most important adaptations by which living organisms (and especially mammals) could prepare their various systems for the coming season. In addition, following the changes in the ratio between *photophase* and *scotophase*, we can anticipate the coming season: When the days shorten we start to anticipate winter whereas when days become longer we anticipate summer.

The ultimate acclimatization of our thermoregulatory system is also entrained by changes in the ambient temperature (Haim 1982). An interesting rhythm in humans which is entrained to the lunar month is the menstrual cycle of about 28–29 days. Why the rhythms in humans are different is still an open question. One possibility is that full moon light can be a signal for sexual activity when males are assured that females are ovulating.

Chapter 4
Biological Clock and its Entrainment by Photoperiod

Abstract As we know today, our biological master clock is located in the supperachiasmatic nuclei (SCN) of the hypothalamus and is entrained mainly by light/dark (L/D) cycles; it also acts as our internal calendar. The modus operandi of our biological clock was considered, for many years, as a black box. Our understanding of the biological clock functioning came from experiments in which subjects were exposed to light during the subjective night, which resulted in a phase shift, while such a shift was not detected when exposure to light took place during the subjective day. As we also know today, the molecular basis of the biological clock is based on interactions between various clock genes paired to create hetero-dimmers and are templates for proteins production that interact with genes by positive and negative feedback loops.

Keywords Biological clock · Pineal gland · Supperachiasmatic nuclei · Melatonin (MLT) · Circadian rhythms · Entrainment by photoperiod · Phase shift response curve · Subjective day · Subjective night · Free running rhythms · Clock genes · Feedback loops

Chronos is the personification of Time, who is often depicted as man turning the Zodiac Wheel.

Greek mythology

Biological Clock as the Main Timekeeper

As we discussed in Chap. 3, humans, like all other terrestrial organisms, have an internal biological clock that also acts as a calendar. While light/dark (L/D) cycles are the main time keepers (*Zeitgeber*, in German), for the entrainment of the clock, changes in the photoperiod are the basis for the prediction of seasonality. Mammals, like other organisms living outside the tropics, use photoperiod changes to

A. Haim and B. A. Portnov, *Light Pollution as a New Risk Factor for Human Breast and Prostate Cancers*, DOI: 10.1007/978-94-007-6220-6_4,

time their reproductive activity (Bronson 1989) as well as to maintain their regulatory abilities (Haim and Fourie 1980; Heldmaier et al. 1981; Haim 1982).

While looking at various natural processes, it sometimes seems surprising to see that plants bloom when their pollinators are active, or, for hibernators, to see their arousal from hibernation when the ice is melted and sun is shining. Similarly, in humans, we can measure daily changes and discover, for instance, that body temperature tends to increase during day hours (even if we are not active) reaching its maximal values in the evening, while at night it drops to its minimal values; body temperature starts rising from its lowest levels when we are sleeping or inactive.

While in Chap. 3 we discussed L/D cycles for the entrainment of our biological clock, in this chapter, we will try to explain where the biological clock is located in mammals, in general, and in humans, in particular, and how it operates on various levels.

Biological Clock and its Functioning

The mechanism of our biological clock was considered to be a "black box" for many years. Little was known about where it is located and how it operates. The clock, which was also known as the "central oscillator" or "pacemaker," was considered a mystery until very recently. Only during the 1980s researchers started to gain information on its location and functioning.

The results of many experiments carried out on circadian rhythms of the Syrian hamster *Mesocricetus auratus*, together with results from phase shift experiments provided a good explanation about the clock abilities to deal with photic and other environmental signals that may entrain it.

Since it is the master clock, it is linked to other biological clocks of lower hierarchies. Bearing in mind that every cell may have its own clock and organs may have peripheral clocks, it would be reasonable to ask: What are the relations between them?

Results of many studies revealed that unicellular algae are able to distinguish between scotophase and photophase, with that ability being attributed, as we previously mentioned, to a simple photoreceptor that has been retained by different organs. It is also considered to be the basic function of the human eye, which entrains the organism's temporal organization to environmental conditions, mainly in predictable environments.

Location and Characteristics of the "Master Clock"

It is clear today that the "master clock" in mammals is located in the Suprachiasmatic Nuclei (SCN) of the anterior hypothalamus (Moore and Eichler 1972; Stephan and Zucker 1972). The critical role of these neurons in the temporal organization of physiological and behavioral daily rhythms of mammals was revealed, among others, by SCN lesions after which animals lost their rhythmicity. Thus in the free ranging Antelope ground squirrel, SCN lesions increased their predation by predators, relatively to intact squirrels (DeCoursey 1986).

Because the "master clock" is a regulatory organ, it is not surprising to find it in the Central Nervous System (CNS). As we know today, the SCN's main pacemaker is a paired neuron cluster above the optic nerve cross (chiasm), containing between 8,000–10,000 neurons which show afferent and efferent neural connections thus having the ability for information input. It is also connected to regulatory centers thus maintaining the daily rhythms or seasonality function. The main input to the SCN comes from the Retino-Hypothalamic Tract (RHT) which receives its stimulus from the bipolar ganglions which contain the protein melanopsin and are known as the non-image forming photoreceptors (NIFP).

In reptiles and birds, as well as in other non-mammalian vertebrates, the pineal gland contains photoreceptors and can directly detect the light and dark period. In the case of mammals this ability was lost presumably during their evolution as subterranean creatures in the form of *Therapsidts*.

The SCN is connected to the pineal gland in two ways: the first one is neural through mainly the L/D information for activating the pineal-cells to produce MLT from serotonin (for more information, see Chap. 6), while the MLT produced as a neuro-hormone and released into the blood system is picked by MLT-receptors distributed on the SCN cells. As a result, the SCN is exposed to neural and neuro-hormonal signals (Nelson 2005).

From the early 1980s on, it was nearly universally accepted that a master clock shows the following important property: on a genetically level in mammals, it is mainly entrained by a photic *Zeitgeber* (time-giver), relatively unaffected by temperature, chemical signals and behavioral feedback. Most information on the entrainment abilities of the master clock comes from the results of experiments dealing with phase shift responses carried out on rodents kept under Free Running Rhythms (FRR) under complete dark (D/D) conditions, constant ambient temperature and no feeding schedules or other *Zeitgeber*.

The first experiments in which the rodents were exposed to LAN and Light Interference (LI) and the responses to such an exposure were measured by the shift response curves of locomotor activity or body temperature daily rhythms. The final evidence for the location of the master clock in the SCN came from a study carried out on Syrian hamsters *Mesocricetus auritus* in the laboratory of Professor Mike Mennaker where three types of hamsters with different periods (20, 22 and 24 h) were kept under FRR conditions (Ralph and Mennaker 1988).

The hamsters with a 22 h period were hybrids obtained from cross breeding between 24 and 20 h types. Breeding among the heterozygote resulted in homozygote and heterozygote types, thus proving experimentally that the biological clock properties in mammals are inherited, meaning they are genetically transferred from one generation to the other as any other property of our body. This model served also for an important experiment in which the SCN was removed from individuals of 24 h type and implanted into a 20 h type (donors and

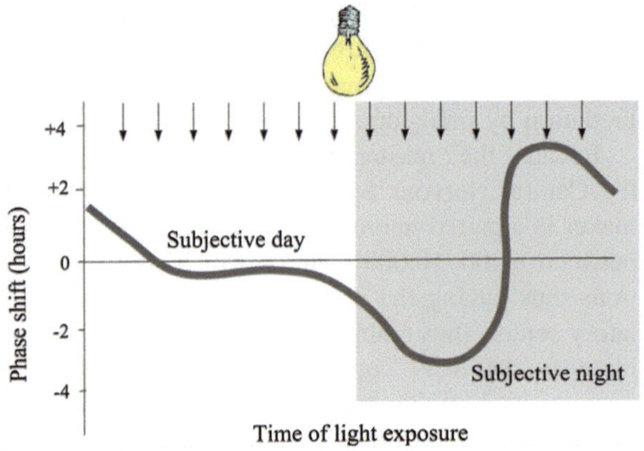

Fig. 4.1 Time of exposure to light stimulus resulting in a phase shift response

acceptors) in a reciprocal manner. The removal of the SCN from a donor (24 h) and its implantation into an acceptor (20 h) revealed that the donors SCN functioned in the acceptor and expressed its genetic background. As experiments were reciprocal and in both cases the expression of the rhythm period was of the donor type, made it clear that the SCN is the master clock.

The Phase Shift Response Curve in Animals and Humans

The results of several studies (cf. inter alia Aschoff and Pohl 1978; Moore-Ede et al. 1982; Wever 1989) indicated that light exposure administrated during the subjective day does not cause any changes in the monitored daily rhythms. However, light given during the dark phase, or close to it, affects the daily rhythms, which is known as the "phase shift effect" or "phase shift response" (see Fig. 4.1).

When the light is given close to the beginning of the dark phase, the next daily rhythm is delayed. Under light exposure, this shift goes on until the middle of the subjective dark phase. An advance in the measured daily rhythm starts at the second half of the subjective night, with the maximal response at about two hours before the subjective day.

Closer to sunrise, the advanced response declines to lower values, and, as mentioned above, no shift is recorded during the subjective day. The phase response curves obtained for many species (such as that featured in Fig. 4.1), enabled chronobiologists to learn about the properties of the biological clock and its photic responses.

Other cues, such as olfactory or noise, could have been and were used for stimulating the system, but no doubt that the response to the photic signal is the most significant one. Furthermore, entrainment of endogenous rhythms to a timekeeper is the result of daily shifts which correct the difference between the period of the central oscillator and the period of the external time-giver (*Zeitgeber*).

Phase shift responses were also studied in humans with similar protocols used for animals (Minors and Waterhouse 1981). The results were quite similar to those obtained from experimental rodents where phase advances and phase delays of the circadian core temperature rhythm were obtained with a single bright light pulse. The main difference between humans and other experimental species relates to the light intensity required to achieve a phase shift response in humans. In humans, it was found to be much higher than in rodents. The authors concluded that the phase response curve is essential for temporal homeostasis in humans and the basis mechanism for entrainment to the solar day.

In the other side of the Atlantic, scientists from various institutes in Boston, MS studied phase shift response as a marker for the sensitivity of human circadian pacemaker, to nocturnal illumination (Zeisler et al. 2000). Results of these studies showed that an exposure of the human vision system to room light early in the morning, can advance timing of the human circadian oscillator. The aim was to study "the dose–response relationship of the human circadian pacemaker to late evening light of dim to moderated intensity." The MLT suppressive effect of light exposure was assessed from MLT plasma levels. The results obtained in the study revealed that humans are highly responsive to light exposure during the early subjective night.

As in other mammals, the phase resetting response to light and the acute suppressive effects of light on plasma MLT appears to follow a logistic dose–response curve. An important result emerging from this research was that in half of the maximal phase-delaying response cases, it was due to exposure to a single episode of evening bright light that can be obtained with dim room light which is one percent of the light used in other experiments. From these results it was concluded that even small changes in ordinary light exposure during the late evening hours can significantly both reduce plasma MLT concentrations and entrain the phase of the human circadian pacemaker.

In regards to humans, we need to keep in mind that they cannot be acclimated for a long period of dark day—dark night (D/D) conditions as rodents. In a study by Khalsa and colleagues (Khalsa et al. 2003), phase response curves were studied in 21 healthy entrained subjects kept under dim light conditions in constant routines in a highly controlled laboratory environment. The subjects were exposed to high intensity light at different points of the circadian cycle in different subject (*ibid.*). Plasma MLT levels were then used as the marker for daily MLT rhythm being followed and compared between pre- and post-stimulus periods. The study was carried out in the Women's Hospital, of the Harvard Medical School.

The results showed that MLT phase delays occurred when the exposure to the light stimulus was before the critical phase (minimal core temperature), while phase advances took place when exposure to light stimulus was after the critical minimal core temperature, while no phase shift in MLT rhythm took place at the critical phase. Interestingly about ten years ago, authors concluded that the results have "practical relevance for the design of light exposure interventions for setting the circadian pace-marker in conditions such as shift work, sleep disorders with a circadian component, and "jet lag". LAN and light pollution defiantly were not an issue in the medical field. The data also suggest that exposure at day time to outdoor as well as to artificial indoor illumination during photophase and darkness during scotophase contributes significantly to timing and entrainment of the circadian pacemaker.

Functioning and Mechanism of the SCN

Knowledge obtained from phase shift curves and from re-movement of the SCN (that is, replacing it with that of a donor and showing that the functioning rhythm is not of the acceptor but rather of the donor) opened new research opportunities for studying the mechanism of the biological clocks' function from the behavioral and physiological levels to the molecular level, where the ultimate goal would be to identify the genes involved and the way in which the genetic code is transferred to regulate various daily rhythms.

In order to follow various SCN functions, different tools were applied, including chemicals that can enhance or inhibit processes occurring on the cellular level or following the metabolism processes taking place on a cellular level. For instance, in order to study rhythmicity patterns of the SCN neurons in rats, the researchers used 2-deoxyglucose (2DG) and autoradiography. As was shown, the daily rhythm of glucose uptake by the SCN neuron cells took place at the day time, bearing in mind that the rat is a nocturnal species (Schwartz and Gainer 1977, Aujard et al. 2001).

Considering that in humans such experiments would be difficult to carry out, how can we really know whether or not the SCN is our main pacemaker?

The information we have is mainly based on clinical reports coming from patients with brain injuries. It was noted that injury of the brain in the anterior parts of the hypothalamus close to the optic chiasm was associated with the disruption of daily rhythms related to body temperature and sleep/awake cycles (Schwartz et al. 1986, Cohen and Albers 1991).

The relations between phase response curves in the SCN and adjacent hypothalamus were also studied in the diurnal murid rodent, *Arvicanthis niloticus* (Mahoney et al. 2001). In their study, the researchers used light induced Fos-immunereactivity in the SCN.

In experiments carried out on rodents, an addition of a running wheel to the cage resulted in some subjects to become nocturnal thus providing a model in which subjects of the same species, but with different temporal organization, could be compared. SCN lesions resulted in the loss of rhythmicity under DD-conditions as well as under DL-conditions.

The results of similar studies also revealed that activity patterns in both diurnal and nocturnal animals were less affected by light given during the middle of the subjective day although both types were affected by light given during the subjective night. While light was given in the early part of the night, it caused a phase delay while when it was given in the late part of the night, the shift resulted in a phase advance. In regards to Fos expression upon exposure to light at different times, revealed that light induced an increase in SCN Fos immune-reactivity during the subjective night only, similarly to the response of activity patterns. This effect was mainly noted in the ventral parts of the SCN, an area characterized in a high concentration of retinal inputs. The authors conclude that although *A. niloticus* is a diurnal species it appears to be similar in their phase response curve to nocturnal rodent species in regards to activity and light-induced Fos expression in the SCN.

An interesting point raised by Mahoney et al. (2001) relates to the gaiting mechanism which determines the response to light pulses. As the authors of this survey suggested, these mechanisms may be different from the mechanism which determines the relationship between the active phase of the rhythm and the Zeitgeber (in this case L/D cycles). This mechanism remains unknown although it may involve differences within subset of SCN cells, which are not identified in cells around the SCN that receive their input from the SCN.

Differentiation in SCN regions and cells is also well documented (Hastings et al. 2003). By recordings from individual cells the authors of this study showed that not all cells have the same pacemaker abilities and each cell has its own specific rhythm. This reminds the situation in the heart in which the fastest cells become the pacemaker. Furthermore, results of *in vitro* studies on single SCN cells from rodents, in which electrical activity was measured, revealed that different individual cells had different circadian periods and it seemed as if the period generated by the SCN is an average of the different periods generated by the different cells (Welsh et al. 1995; Hertzog et al. 1998).

The neurotransmitters known to be involved in the connection between the SCN neuron cells are: Arginine Vasopresin (AVP) Vaso-active Intestinal Peptide (VIP) and Gamma-aminobuteryc acid (GABA). GABA administration given at night had an inhibitory effect while at day time it had an excitatory effect (Wagner et al. 1997).

In rats maintained under light/dark cycles it was noted that AVP is secreted at a higher rate during day time relatively to the dark period, this pattern is kept when the rats were acclimated to LL-conditions. This neurotransmitter (AVP) is presented in the dorsomedial parts of the SCN while VIP is mainly abounded in the ventrolateral parts of the SCN. An interesting result in regards to VIP was that an *in vivo* treatment with VIP resulted in phase shifts similar to those caused by light pulses (Piggins et al. 1995). From the different neurotransmitter types discovered in the SCN it may be assumed that this neural structure is differentiated and the different parts may play different roles.

The Molecular Basis for the Master Clock Functioning

Much attention was given to the clock genes of the SCN in order to understand the molecular mechanism of rhythmicity starting with the classical genetic model, the fruit fly *Drosophila*, where the gene *Per* (Period) was first described from mutagenic flies with shorter or longer periods and from flies that became arrhythmic due to a mutagenic treatment. Although the gene and its Protein PER were identified several decays ago (Konopka and Benzer 1971), it took quite a long time to understand how circadian rhythms are regulated by genes. Today we know that the genetic regulation is not simple and it involves several genes. Furthermore, we know that they are differences in the genes that regulate this process in different groups of organisms but the principle will describe here (although they are of different identified genes) seems to be common to *Drosophila* (Young 2000) and mammals SCN cells (Hastings et al. 2003).

The regulation can be described as "molecular feedback loops" where in mammals (mice) the genes and their proteins, which are involved, are known and some of them differ from those in the fruit fly. The circadian clock genes *Per* and *Tim* (Timeless) exist in the fly while in the mice *Cry* (Cryptochrom) was discovered (Reppert and weaver 2002, Roenenberg and Merrow 2001) and these genes can have variations. The knowledge we have today in regards to the mammalian genes controlling circadian rhythms is that these genes show gene loops, which have positive and negative feedback loops involving pairs of genes and their proteins. These relations between the genes in the nucleus and their proteins in the cytoplasm constitute the type of the feedback loop. A pair of genes produces hetero-dimmer in the nucleus and their proteins appear in the cytoplasm, which together initiate the activation of another pair hetero-dimer of clock genes (positive feedback).

The proteins produced suppress the activation of the second pair of genes and this is considered as a negative feedback which will activate once again the first pair of clock genes. The known clock genes in rodents include: *Clock* and *Bmal* which produce together a hetero-dimmer (positive feedback) which can include one type of *Per* (out of three types of *Per* genes) and one of the two types of *Cry* (cryptochrom). The PER and CRY move into the nucleus where CRY interacts with the hetero-dimmer of *Per*/*Bmal* to prevent further transcription, thus acting as a negative feedback (Reppert and Weaver 2002). As in the case of *Drosophila*, much of the research was depended on the discovery of mutants in clock genes. Interestingly, in rodents it was noted that expression of *Per1* and *Per2* increased as a response to light exposure during the subjective night, such a response was not revealed during the subjective day (Sherman et al. 1997), showing the connection between environmental illumination conditions and gene activation (Fig. 4.2).

Entrainment of the Biological Clock in the Real World

Bearing in mind that in the real world we do not experience FRR conditions, our endogenous pacemaker is adapted to entrainment by L/D cycles on a daily basis and changes in photoperiod on an annual basis. A precondition for our biological clock to anticipate the coming environmental changes, or to function as a regulator for temporal organization, is due to its adaptation to a certain location on our

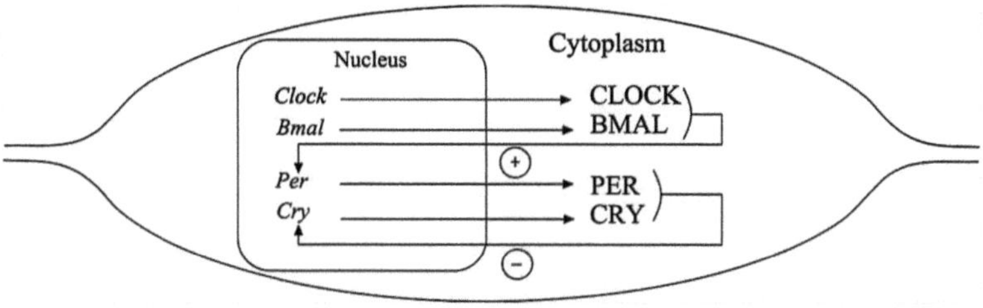

Fig. 4.2 The mammalian gene and protein loops as the molecular basis for the biological clock operation. (Note the existence of positive and negative feedback loops—see text for explanation)

Planet, which can be measured e.g., by its latitude and longitude. Under natural conditions, the move from light to dark and vice versa is gradual in regards to changes in natural light intensity and changes in the photoperiod, thus helping to prepare our physiological and immune systems to the coming season.

Modern life is more complicated, and, at least, two major inventions changed dramatically lives of many people on our Planet by affecting the functions of our pacemaker, as our main temporal organizer. These main inventions are the incandesced bulb, in which the electrical energy is converted into illumination (a topic we are dealing with in a separate chapter of this book), and, the second is the jet engine which enables us to move from one place to another in times which are shorter than the rotation of our Planet around its axis.

Under these conditions, the entrainment of our biological clocks to the natural *Zeitgeber* becomes problematic. Disruption of our temporal organization results in disorganization (or dis-synchronization) of our physiological and immune systems resulting, in turn, in pathological consequences about which we have become aware in the past decades, including the loss of harmony in different daily rhythms. The increase of nighttime light intensity worldwide (especially of short wave length illumination) brings people all over the world closer to environmental conditions similar to those of shift workers.

Chapter 5
Light at Night (LAN) Exposure and its Potential Effects on Daily Rhythms and Seasonal Disruptions

Abstract Light/dark (L/D) cycles are the environmental cue for the entrainment of our circadian rhythms while changes in photoperiod are an important environmental cue for seasonal acclimatization of different physiological systems as well as for immune system functioning. The neuro-hormone melatonin (MLT) plays a major role in this process as it indicates the dark period of the 24 h cycle. Therefore the disruption of MLT secretion due to Light at Night (LAN) exposure may have a negative impact on daily rhythms and seasonality of our physiological and immune systems. Supposedly the suppression of pineal MLT production has a direct effect on cells which contain MLT receptors.

Keywords LAN · Seasonality · Photoperiod · Acclimatization · MLT receptors

We can easily forgive a child who is afraid of the dark; the real tragedy of life is when men are afraid of the light.

Plato

Entrainment of the Biological Clock and Calendar

As we already discussed in the previous chapter, our daily rhythms are an expression of the entrainment of the endogenous biological clock (the master oscillator or our time pacemaker), located in the SCN, to exogenous L/D cycles. Other types of external time-keepers (*Zeitgeber*), such as lunar cycles, entrain tidal rhythms in organisms inhabiting the tidal zoon. In all mammals studied so far, under the dark conditions during the night, the pineal gland produces and secrets the neuro-hormone Melatonin (MLT), which, among its other functions, is transferring the dark period signal to cells, tissues and organs of our body with even the SCN containing MLT receptors.

Exposure to LAN suppresses the MLT production (Armstrong et al. 1989) and therefore, it should be considered as a potential disrupter of daily rhythms. Light

A. Haim and B. A. Portnov, *Light Pollution as a New Risk Factor for Human Breast and Prostate Cancers*, DOI: 10.1007/978-94-007-6220-6_5,

given at subjective night time to hamsters or mice and rats kept under Free Running Rhythms (FRR), with no exposure to any other *Zeitgeber*, caused what we define as the *phase shift response*, and such curves are known for different rodent species, as we already discussed elsewhere in this book (see e.g., Chap. 4). Interestingly, a difference in response to LAN exposure was noted when LAN was given in the early hours of the dark period or close to the midnight or in the second half of the dark period. The main conclusion from these experiments was thus that light given in the dark period under FRR conditions can cause a phase shift response, thus affecting daily rhythms, such as activity, body temperature, as well as pineal MLT production and secretion (Moore-Ede et al. 1982).

Most of urban populations worldwide are being exposed today to artificial illumination with a growing exposure to short wavelength illumination, which is becoming the most common source of illumination in public places and in indoor spaces. When we sit in front of a TV screen, use our cellular phone or PC at night, or are simply exposed to indoor illumination, MLT production could be suppressed and this should be a matter of concern in regards to our health and negative impacts of LAN on it.

In addition to the daily rhythms of hormones secretion, the physiological and behavioral changes promise the best response of our body to cope with our environment for instance by maximal alertness at the time of foraging. Such an adjustment was achieved through natural selection and can be considered as an adaptation of human and/or other organisms at a certain locality on our Planet.

Bearing in mind that humans are diurnal organisms, the day time is devoted to activities such as foraging and reproduction, while the dark period of the 24 h cycle is devoted to sleep. Hence the secretion of hormones occurs in a certain order, we may not be fully aware yet of all the consequences of, for instance, the activation of receptors of a certain hormone by a hormone secreted earlier in time. Timing of endogenous functions is critical in a changing environment and if a predator is active when its' pray is hiding in a safe place, most probably he will stay hungry and will not compensate for the energy spent during foraging. Or, if a pollinator is active at the time when the flowers pollinated by it are closed, it will not be able to pollinate and will also have no nectar which is its energy main source.

If timing is so important, how comes that we, humans, underestimate the importance of natural illumination for entraining our biological clock and keeping us in harmony with temporal and spatial variables of our environment, which, in many cases, are separated from us artificially?

Humans have tried to extend light hours by various ways throughout the entire history of mankind (see Chap. 7). Beyond any doubt, humans' "victory" over "nighttime darkness" came about with the discovery of the incandesced bulb in which electrical energy is transferred into illumination and reaches light intensities which humans had never faced before during nighttime, thus losing an important natural signal for the entrainment of our biological clock and as a consequence the adjusted functions of physiological and immune systems.

The increased light intensity seems to become a recently recognized source of pollution and the American Medical Association (AMA), the largest association of medical doctors worldwide, which has recently passed a resolution stipulating that LAN is a source of environmental pollution which interferes with our daily rhythms. This is a very important decision coming after that of the subcommittee of the World Health Organization (WHO) which declared that night shift work is potentially carcinogenic for humans (Straif et al. 2007). In both cases, it seems that the disruption of our temporal organization is considered as an event with negative outcomes for our health.

Seasonal Acclimatization in Animal Models: The Importance of Photoperiod

As a reflection of seasonality in winter mammals, facing low ambient temperatures, increase minimal heat production and decrease heat dissipation in order to avoid *hypothermia*, while in summer they decrease minimal resting heat production and dissipate heat more efficiently, to avoid *hyperthermia*. Results of many studies carried out on various rodent species revealed that acclimation to short day (SD) photoperiod is an initial cue for winter acclimatization of the thermoregulatory system (see *inter alia* Haim and Zisapel 1995).

Some aspects of the thermoregulatory system, based on morphological data of the South African striped mouse *Rhabdomys pomilio*, were studied by Coetzee (1970). This mouse is a diurnal and crepuscular species, widely distributed in Southern Africa and, in an experiment carried out on this species it was shown that moving 12L:12D acclimated mice from 25 to 6 °C, resulted in the death of most subjects with no acquired resistance to low ambient temperatures. However, moving of mice first to a photoperiod regime of a short day, 8L:16D under 25 °C for 2 weeks (that is, to decreasing photoperiod hours) resulted in increased resistance of all individuals to low ambient temperatures (Haim and Fourie 1980b). Therefore, it was concluded that for *R. pomilio*, changes in photoperiod, are the initial environmental cue for seasonal acclimatization of the thermoregulatory system which takes place before the decrease in ambient temperature.

A MLT treatment of 12L:12D-acclimated *R. pumilio* kept at 25 °C also resulted in increasing resistance upon exposure to cold and they did not become *hypothermic* (Haim and Fourie 1982). Therefore, it could be assumed that acclimation to short photoperiod results in an increase in pineal MLT production due to an extension of the dark period.

The social vole *Microtus socialis* is a pest rodent, common to agricultural fields in Israel. Various aspects of its physiology were studied, including its reproduction patterns, where its main breeding activity takes place under short days and long nights (Brandes et al. 2004). Bearing in mind that the Levant is the most southern part of its geographical distribution, the agricultural fields in Israel are the southernmost border of its distribution. However, this species is abundant in the steppe ecosystems on the Golan Heights from where breakouts were described.

Reproducing during the short photoperiod when temperatures are cool but fresh food and water are plenty can be an advantage. The question the researchers asked many years ago was as follows: *Do the voles use photoperiod changes as a cue for seasonal acclimatization of their thermoregulatory and reproductive systems activation*?

Results of a study carried out under laboratory conditions revealed that SD-acclimations of the voles improved their resistance to low ambient temperatures helping to maintain their T_b relatively high with a lower increase in metabolic rates. The results suggest that an increase in insulation properties, compared to LD-acclimated voles which upon exposure to cold could not maintain T_b to the same levels as SD-mice and exhibit relatively high metabolic rates. However, under LD-acclimation voles could cope better with high ambient temperatures compared to SD-acclimated ones (Banin et al. 1994).

The 1977 autumn vole breakout in *alfalfa* fields, in the northern parts of the Jordan Rift Valley, brought us to suggest the use of photoperiod disruption by LAN to control vole population size. If most of the breeding takes place under short days, LAN should disrupt reproduction. In a field experiment carried out using specially delineated enclosures, LAN resulted not only in reducing reproduction but also in the death of the LAN exposed voles, while in the control enclosure the vole population almost tripled in size (Haim et al. 2001, 2004).

The results of laboratory studies further revealed that SD-voles acclimated to LAN could not maintain their heat production abilities upon exposure to cold. Under acclimation of SD-voles to LAN, not only daily rhythms and reproduction were disrupted, but so was the seasonality of their thermoregulatory mechanisms. In both cases, the daily rhythms and seasonal acclimatization could thus be connected to the suppression of MLT production upon exposure to LAN.

An important mechanism for heat production in small mammals, such as rodents, is non-shivering thermo-genesis (NST), when heat is produced by catabolism which does not involve muscle contraction (shivering). This mechanism was found to be effective and responding quickly in cold acclimated rodents compared with normo-thermal acclimated ones. It was noted that a sympathetic stimulus activates different tissues by releasing noradrenalin (NA). In laboratory experiments, exogenous NA was used for activating NST and the response was detected in order of minutes (Jansky 1973). As the NA dose is mass-depended, different results obtained under different acclimation conditions may hint on the role of NA receptors. Bob Lynch was the first who showed that acclimation of *Peromyscus leucopus* to short day resulted in an increase of NST levels, as if the SD-individuals of *P. leucopus* were cold acclimated (Lynch 1970). Similar results were obtained latter on for different rodent species (Haim 1982; Haim and Yahav 1982) and an interesting question asked in regards to this response was: *What is the mechanism behind the NST response to the increase in the dark hours*?

One possibility could be that the extended secretion of MLT may affect the receptors to noradrenalin in brown adipose tissue (BAT) or other heat producing tissues as the stratified muscles. This change can be in number of receptors or their affinity to NA and these topics to the best of our knowledge were not studied in relation to LAN. Although results of a NST study in which *M. socialis* were SD or LD-acclimated revealed high and low NST capacity respectively and are considered as the initial environmental cue for seasonal acclimatization, indicating that LAN can interfere with seasonality of physiological mechanisms presumably among others by affecting affinity of receptors.

An early study by Haim et al. (1983) on the thermoregulatory abilities of the "Blind" mole rat *Spalax eherenbergi* to respond to photoperiod changes indicated the existence of a more complicated retina and this finding led to the discovery of the non-image forming photoreceptors (NIFP) which are essential for the entrainment of the circadian rhythm. Results from a recent study carried out in the Chronobiology research center at the University of Haifa revealed that the "Blind" mole rat responds to LAN (Zubidat et al. 2011). The results of this study revealed that exposure to short wave length (blue) or long wave length (red) illumination for half an hour close to mid night, of SD-acclimated "blind" mole rats acclimated to one or three weeks of LAN exposure, resulted in a suppression of MLT production, thus supporting the idea that LAN suppresses MLT production and that this species is red-shifted.

Seasonality in the Functioning of the Immune System

As LAN is associated with the suppression of pineal MLT production, this association may affect the immune system. The fact that lymphocytes contain MLT receptors (Nelson and Drazen 1999) puts the immune system on the "same page" with other systems that show seasonality which may be explained by levels and duration of circulating MLT in the plasma. Syrian hamsters treated with MLT, or hamsters acclimated to SD-conditions were found to increase MLT levels as well as total splenic lymphocyte counts, mass and numbers of macrophage (Vaughan et al. 1987). Day–night variations in MLT binding to spleen membrane were reported and MLT suppression by exposure to LAN could suppress immunostimulatory properties, while L:L conditions inhibits T cell autoimmunity as a result of MLT suppression (Refii-El-Idrissi et al. 1996).

Results of a study on humans kept for 40 h awake revealed significant changes in various variables of the immune system functions, including a decrease in natural killer cell activity (Moldofsky et al. 1989). As Navara and Nelson (2007) concluded, exposure to LAN can significantly modulate immune function, leading to large-scale medical implications. Thus, the response of a mammal to LAN can be a direct result of MLT suppression or an indirect result stemming from a malfunctioning of the immune system.

Results of another recent study carried out in the Chronobiology research center at the University of Haifa on the golden spiny mouse *Acomys russatus* as an animal model, revealed differences in the immune response to KLH injection. While in

SD-acclimated mice the response was fast, it was found to be slower in LD-acclimated mice. However, SD-mice exposed to LAN showed a response similar to that of LD-mice. A major difference among the groups was in regards to the atypical white blood cells (that is, cells which are not differentiated). While these cell counts were high in LD-acclimated mice they were low in SD-mice. Acclimation of the latter to LAN resulted in an increase in atypical cells similar to those of LD-mice. MLT treatment to SD+LAN acclimated-mice increased differentiation, thus reducing values of atypical white cells. These results are in agreement with those reported in the literature and suggesting an adverse effect of LAN on the immune system.

Chapter 6
Melatonin: "Hormone of Darkness" and a "Jack of all Traits"

Abstract The pineal neuro-hormone Melatonin (MLT) is involved in the regulation and functioning of various processes in the human body. Among other functions it plays a significant role in the transferring of the dark period signal on a daily basis while its levels and duration of its secretion are the basis for seasonality. As a result the suppression of MLT production by LAN exposure affects not only various daily rhythms but also seasonality of our thermoregulatory system being responsible for heat production on the one hand for heat dissipation on the other. The disruption of the MLT production also affects our immune system due to the fact that receptors of this hormone are distributed on white blood cell. Because MLT is an anti-oxidant and anti-oncogenic agent its reduction from exposure to LAN can result in increased breast prostate cancer (BC&PC) risks through a more rapid proliferation of cancerous cells or/and in an indirect way by a mal-functioning of the immune system. As MLT suppression depends not only on light intensity but also on light wavelength with short wavelength illumination being very effective in MLT suppression. The increase of such illumination worldwide under the "environmentally friendly illumination" paradigm should thus become a concern for health authorities worldwide.

Keywords Pineal · LAN · MLT suppression · Breast and prostate cancers (BC&PC) · Thermoregulatory function · Immune system

No matter how fast light travels, it finds the darkness has always got there first, and is waiting for it.

Terry Pratchett

Melatonin Hormone: Discovery and Chemical Properties

MLT (*N-acetyl-methoxytryptamine*) is a pleiotropic neuro-hormone produced and secreted by the pineal gland during the dark phase of the 24 h cycle, mostly under dark conditions (Reiter and Fraschini 1969; Blask et al. 2005). MLT secretion is

A. Haim and B. A. Portnov, *Light Pollution as a New Risk Factor for Human Breast and Prostate Cancers*, DOI: 10.1007/978-94-007-6220-6_6,

inhibited by light, even when the duration of light interference is short and of relatively low intensity (Brainard et al. 1997). Moreover, MLT inhibition in humans by light was demonstrated to be wavelength dependent, with shorter wavelength (greener and bluer) illumination being more effective in MLT suppression than yellow–red illumination of longer wavelengths (Cajochen et al. 2005).

Altogether, the exposure to LAN through the disruption of MLT production (affecting daily rhythms thereby), interferes with our temporal organization, possibly resulting in cellular malfunction, and promoting the abnormal proliferation of modified cells.

MLT was first discovered by McCord and Allen (1917) who showed that an extract of bovine pineal gland given to frogs resulted in the bleaching of their integument. Some 40 years later, Aaron Lerner, a dermatologist from the University of Minnesota, together with his coworkers from Yale University, succeeded to identify the chemical extracted from the pineal gland and named it *melatonin* (Lerner et al. 1958). Since its discovery, MLT existence was noted in all studied organisms, ranging from unicellular (such as *algae*), to all plants, and invertebrates and all vertebrates in the animal kingdom (Reiter et al. 2003; Hardeland and Poeggeler 2003; Kohidai et al. 2002).

Results of many studies carried out to date allow us to conclude that MLT plays a major role in many processes and functions of living organisms, and, therefore, can be considered as a "jack of all traits" being, among others, a very effective anti-oxidant and anti-oncogenic agent (Kolar 1997; Poeggeler et al. 2002; Tan et al. 2002; Garicia et al. 2010).

MLT is produced from serotonin, which procurers the amino-acid tryptophan. Serotonin is accumulated during the light hours (photophase) in the pineal gland (pinealocytes) and is transformed into MLT during scotophase in two steps, where, in the first step, the enzyme arylalkylamine N-acetyltransferase (AA-NAT) is involved, while, in the second step, Hydroxyindole-O-methyltransferase (HOMT) is involved. The former enzyme is under control of the Suprachiasmatic nucleus (SCN) and is produced at night, under dark conditions.

MLT is a very lipid-soluble small indole molecule possessing empathetic characteristics to the others. As a result, it can pass very easily through the cell membrane and its receptors are identified on the nucleolus membrane MT1 and MT2 from the G protein-coupled receptor group as well as receptors from the RZRa and RZR/RORbeta family (Hardeland et al. 2006).

Pineal MLT production and secretion increases after dusk and, in human, it picks up between 2:00 and 4:00 h in the morning, while after this period of time, its production slows down and stops about 3 h after sunrise (Reiter 1993).

As we also know today, light interference (LI) disrupt MLT production and secretion. The results of a classical experiment carried out by Cajochen and coauthors (2005) revealed that exposure to LAN of short wavelength 460 nm is highly effective in suppressing pineal MLT production, while under the same intensity and for the same exposure duration, but under the wavelength of 550 nm, such an effect does not occur (see Fig. 6.1). Therefore, for artificial suppression of MLT, such as in Seasonal Affective Disorder (SAD), the use of high intensity short wave length illumination is very efficient (Nelson 2005).

MLT Functions

As we already mentioned, the immune system shows daily rhythms and seasonality. An increase in the mass of the thymus gland was noted in short day-acclimated young voles (Vaughan et al. 1973). Moreover, a correlation was noted between levels of serum MLT and phagocyte activity. The existence of MLT receptors on lymphocytes is another important indicator for interaction between MLT, which directly responds to the temporal changes in our environment (by its levels and secretion duration at night), and the immune system (Guerrero and Reiter 2002). MLT is involved in many other functions in the human body.

Among its other functions, MLT transfers the dark signal to all body cells and tissues maintaining the original primary function of a photoreceptor production, distinguishing between *photophase* and *scotophase*, while the relations between them (photoperiod) is also transferred by MLT thus signaling seasonality to the cells and tissues.

MLT also regulates the activity of the reproductive system in seasonal breeders, a function which is well documented in many mammal species. For instance, in long day breeders, it suppresses reproduction, while in short day breeders, such as the social vole (*Microtus socialis*) and the mole rat (*Spalax ehernbergii*), it activates the reproductive system. In sheep MLT treatment is used in a commercial

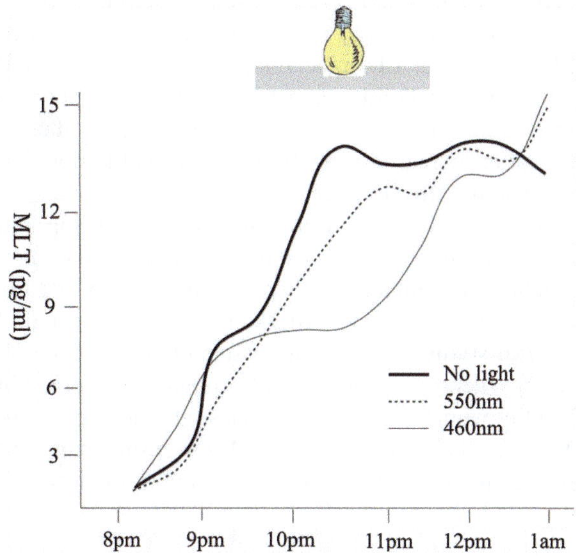

Fig. 6.1 Effects of a 2-h light exposure at 460 and 550 nm on salivary MLT levels (diagrammed using data from Cajochen et al. 2005)

way for breeding twice a year where sheep are treated with MLT administrated under increasing photoperiod after natural delivery. In young humans MLT is known to suppress the development and activation of the reproductive system thus delaying sexual maturity.

In addition, MLT is an efficient anti-oxidant not only in the animal kingdom but also in plants including algae. In plants MLT main role is of a scavenger of free radicals produced in photo-oxidative processes. In mammals, MLT is produced also in the retina, thus acting as an anti-oxidant removing free radicals produced as a result of photo-oxidative processes taking place in the retina upon exposure to light at day time, which can be looked at as a parallel function to photo-oxidation taking place in plants during photosynthesis.

MLT is also known as a direct anti-oncogenic agent for two types of cancers: BC in females and PC in males. In regards to BC, several mechanisms are suggested for its role, including modification of estrogen receptors. MLT treatment to BC cells increases methylation to levels observed before exposure to LAN.

MLT and its Links to BC&PC

In an *in vivo* experiment carried out by Haim and his colleagues (Haim et al. 2010) *Balb* male mice were inoculated *s.c.* with mice prostate cells (TRAMP) and after inoculation they were kept under two different photoperiod regimes namely short and long days (SD-8L:16D and LD-16L:8D respectively). Half of the SD-mice group was interfered with light (450 lux, with a dominant wave length at 469 nm) for 30 min, 7 h after lights went off, while another half of the LD-mice were treated with MLT first by injection and then by adding it to the drinking water given during the dark phase. Under LD-condition tumors were significantly bigger with a higher growth rate comparing with SD-mice thus, suggesting that lower MLT levels in LD-mice relatively to SD-mice. Indeed the results of the MLT treatment to LD-mice and LI to SD-mice support this idea, as growth rates and final tumor size in the former were similar to the values of SD-mice, while LI to the latter increased growth rates and final tumor size were increased significantly compared with SD-mice.

High MLT levels during scotophase are important for regulation of pituitary and ovarian hormones such as estradiol, and also for increasing DNA-repair mechanisms, enhancing thereby the function of pathways which may prevent the development of cancer (Blask et al. 2005; Cos et al. 2000). Furthermore, results of clinical studies revealed an association between MLT levels and metastatic-BC, where a decrease in MLT peak amounts were noted in BC-patients, as compared to healthy women. Studies also showed that larger tumors are associated with lower MLT levels (Cos et al. 2000). These results support the idea that MLT is an anti-oncogenic agent, as known from animal models.

LAN→Pineal MLT Suppression→BC&PC Cells Increased Proliferation Rates

According to Brzezinski (1997) and Brzezinski et al. (2005), MLT synthesis and release occurs in a dose–response-like manner stimulated by darkness and inhibited by light. MLT peak levels normally occur during sleep after midnight. These levels decrease and become minimal when we are exposed to day light. In his paper on LAN, shift-work and BC risk, Hansen (2001b) calls for further exploration of the relationship between exposure to LAN and shift-work, including timing during the night, and cancers that may be MLT depended. Today, some 10 years later, this call is relevant not only for shift workers but also for large urban populations exposed to LAN indoors and outdoors as well as for residents of rural areas, where home illumination, including electronic devices with LED lights become a source of light pollution.

It has been also revealed that high MLT levels at day time result in Seasonal Affective Disorder (SAD) relatively common in human populations of northern America and in northern Europe. This issue is dealt mainly by psychologists and psychiatric physicians, but also by chronobiologists, as a common cure for SAD is exposing the subjects to short wave length illumination which suppresses MLT (Nelson 2005). Once again showing the importance of timing, the reduction of MLT secretion at day time is crucial for our health, while disruption that may emerge from an insufficient amount of light at day time will also have negative health effects.

The importance of exposure to natural day illumination is of great importance for the resetting of MLT production. The short wavelength illumination suppresses MLT production and at the same time our skin produces vitamin D. *If increase in MLT levels is a signal transferred to cells and tissues during the dark period, may high levels of vitamin D become a signal for the light period in addition to the marginal levels of MLT?*

Our modern life can be complicated, as in lower latitudes (for instance, in the Mediterranean basin) we prefer not to be exposed to solar radiation from 10:00 to 16:00 h in summer in order to avoid melanoma. However, many people are not exposed to natural illumination as they leave home for work and come back after sunset. Such behavioral patterns result in the fact that MLT production is not completely suppressed, while vitamin D production is suppressed. As a result of LAN exposure, environmental temporal cues are not transferred correctly to the body cells or to the master clock for adjusting our temporal organization.

Furthermore, the dramatic increase in short wave length illumination under the term of "environmentally friendly illumination" does not result only in the increase of light intensity in general but specifically in the increase of short wavelength illumination, which efficiently suppresses MLT production (Feleci et al. 2011). LED illumination in electronic devices surrounds us and accompanies us even in our sleeping habitat. As a result, children are exposed to such illumination for many hours during the dark period, when they sleep, that is, during the time when MLT is to be produced under dark conditions. As MLT levels are not measured on a regular basis, most of us are unaware of the risks involved.

The existence of the "immune-pineal" axis was recently suggested by several authors (Markus et al. 2007; Markus and Ferreira 2011). As the pineal was discovered to be a target for molecules signaling immune response, it was proposed that mediators of inflammation modulate the synthesis of MLT, where pro-inflammatory mediators inhibit while anti-inflammatory mediate potentiate MLT production. This novel view of interaction between the immune response and the pineal gland may explain various diseases. As Markus and Ferreira (2011, p. 102) conclude, "the pineal gland is not only a transducer of photoperiodic information, but is also a constitutive player in the innate immune response". As we suggest, further research on this potential axis should include BC, PC and other cancer types and their response to LAN.

Finally, in regards to MLT production the following questions are to be asked: *How will the pineal respond to a dark cue at day time? Will MLT be still produced? Is there any threshold in regards to the time of darkness and how will it affect our daily rhythms?*

Answers to these questions will help us to develop a holistic approach helping to facilitate the relations between our biological temporal organization and environmental *Zeitgeber*, namely photoperiod.

Part II
Light Pollution, its Known Health Effects and Impact on Energy Conservation

Chapter 7
Introduction and Spread of Artificial Illumination: A Human History Retrospective

Abstract People knew about electricity for centuries. However the first successful attempt to use electricity for lighting can be credited to Sir Humphrey Davy who discovered in 1801 the incandescence of an energized conductor. Yet the idea of using electricity for lighting "took off" only after the American inventor Thomas Alva Edison developed his deep vacuum incandescent lamp with a carbon cotton filament. Thomas Alva Edison also contributed to the practical use of electricity by installing the first electrical lightning system at Pearl Street in NYC in 1882. Since then both light bulbs electricity production have become relatively cheap and more reliable. As a result of rapid electricity proliferation electric lighting has substituted most traditional lighting sources making human population virtually independent of natural L/D cycles. The full range of implications of this transition for human health are yet to be determined.

Keywords Electricity · Lighting · Incandescence · Light bulbs · Electricity timeline · Electrification rates · Light/dark (L/D) cycles · Health effects

We will make electricity so cheap that only the rich will burn candles.

Thomas Edison, Dec. 1879

The Hanukkah Story

The ancient Jewish historian Josephus Flavius wrote about an early light miracle. According to his narrative, in the year 175 B.C.E., Syrian (Greek) armies invaded Judea. Several years later, in 167 B.C.E., the Second Temple in Jerusalem was looted, cleared of all Jewish artifacts and dedicated to Zeus. Jews were forced to pray to idols and eat pig flesh, both practices explicitly forbidden by the Jewish faith. Traditional Jewish worshipping and ritual sacrifices were also forbidden. Popular discontent was not long to come. It reached its climax in *Mod'iin*, a small village

A. Haim and B. A. Portnov, *Light Pollution as a New Risk Factor for Human Breast and Prostate Cancers*, DOI: 10.1007/978-94-007-6220-6_7,

Fig. 7.1 The Menorah lamp and other articles from the Jewish Temple in Jerusalem in a public display after the destruction of the Second Temple, depicted on the Titus Arch in Rome (ca. 82 C.E)

located a few kilometers west of Jerusalem. The residents of the village were known for their restiveness and often disobeyed the authorities. Eventually, the patience of the Greeks was exhausted and they sent a small group of soldiers to teach the villagers a lesson of obedience.

As the story goes, the soldiers gathered the villagers in the central square, where the villagers were forced to bow to an idol and eat pork. Most of them refused, and only one agreed. *Mattathias the Hasmonean*, the local priest, grabbed a sword and killed the traitor. Then he turned his wrath on the commanding officer of the Greek contingent and killed him as well. Following *Mattathias's* lead, the villagers drew off their swords and knives and quickly killed the rest of the Greeks. The revolt spread quickly across the land, led by *Mattathias* and his five sons—*Yochanan, Shimon, Elazar, Jonathan*, and *Judah*,—who later became known as the *Maccabees (or Maccabim)*, the name used until today for various Israeli sport teams and for Jewish Olympic games known as "Macabia".

The uprising was successful, and in 165 B.C.E. the Temple in Jerusalem was liberated. Since the Temple was used by the Greeks for idol worshipping, it needed to be purified by cleaning of all foreign artifacts and burning sacrificial oil for 8 days. However, to everyone's dismay, only a small jug of uncontaminated oil was found, which could last for 1 day only.

The Maccabees lit the ritual lamp anyway, and, much to everyone's surprise, the small jug of oil lasted for full 8 days. Jews around the world celebrate the event to this very day. Remembering the miracle, they light up a special lamp, known as the *Hanukkiyah,* for 8 nights and days, adding every night an additional candle (Fig. 7.1).

Discovery of Electricity

Another remarkable event in the ancient history also related to light happened some 435 years before the Maccabee rebellion. Around 600 B.C.E., the Greek philosopher, Thales of Miletus made a remarkable discovery. He observed that an amber stone, when rubbed, attracted small light objects. In the present-day terms, the phenomenon discovered by Thales is known as static electricity.

However, it took another 300 years to find out that other substances possess similar electro-static properties. This discovery belongs to another Greek philosopher, Theophrastus of Lesbos (371–287 B.C.E.). Yet, the very word "electricity" was coined by the English astronomer William Gilbert some 2200 years later. The term was derived from "electron," the Greek word for amber.

For centuries, no particular attention was paid to Thales's and Theophrastus's discoveries, which were perceived as artifacts having little practical value. For centuries, studies of static electricity and magnetism were mainly limited by the improvement of compasses used for maritime navigation (Fig. 7.2).

Only in 1657, the British scientist Robert Boyle discovered that electric force can be transmitted through vacuum. Several years later, in 1663, Otto von Guerick built the first electrostatic generator, or an electrostatic machine that produced static electricity by friction. This machine had no practical use, and no explanation was offered to explain its operation. It was demonstrated solely as a curious *objet d'art*.

No application of electricity for lighting was sought then. In the absence of other options for prolonging daylight and enabling various activities after sunset and during dark overcast days, people lit their houses with various primitive means, such as oil lamps, candles, burning wood, coal and kerosene lanterns, whenever available (Fig. 7.3).

Further Developments

It took another 70 years to discover that an electric discharge results from the proximity of positive and negative charges. This remarkable discovery was made in 1733 by the French scientist Francois de Cisternay Du Fay. One decade later, Ewald Georg von Kleist and Pieter van Musschenbroeck from the Leiden University in the Netherlands demonstrated that electricity can be stored. They created the first electric condenser known as the Leiden Jar.

Inspired by these discoveries, Jean-Antoine Nollet built in 1748 the first electroscope, and in 1752, Benjamin Franklin, Thomas-François Dalibard and DeLors published their pioneering studies on electrical properties of lightening. As a practical application of his research, Benjamin Franklin invented in 1753 the lightening rod.

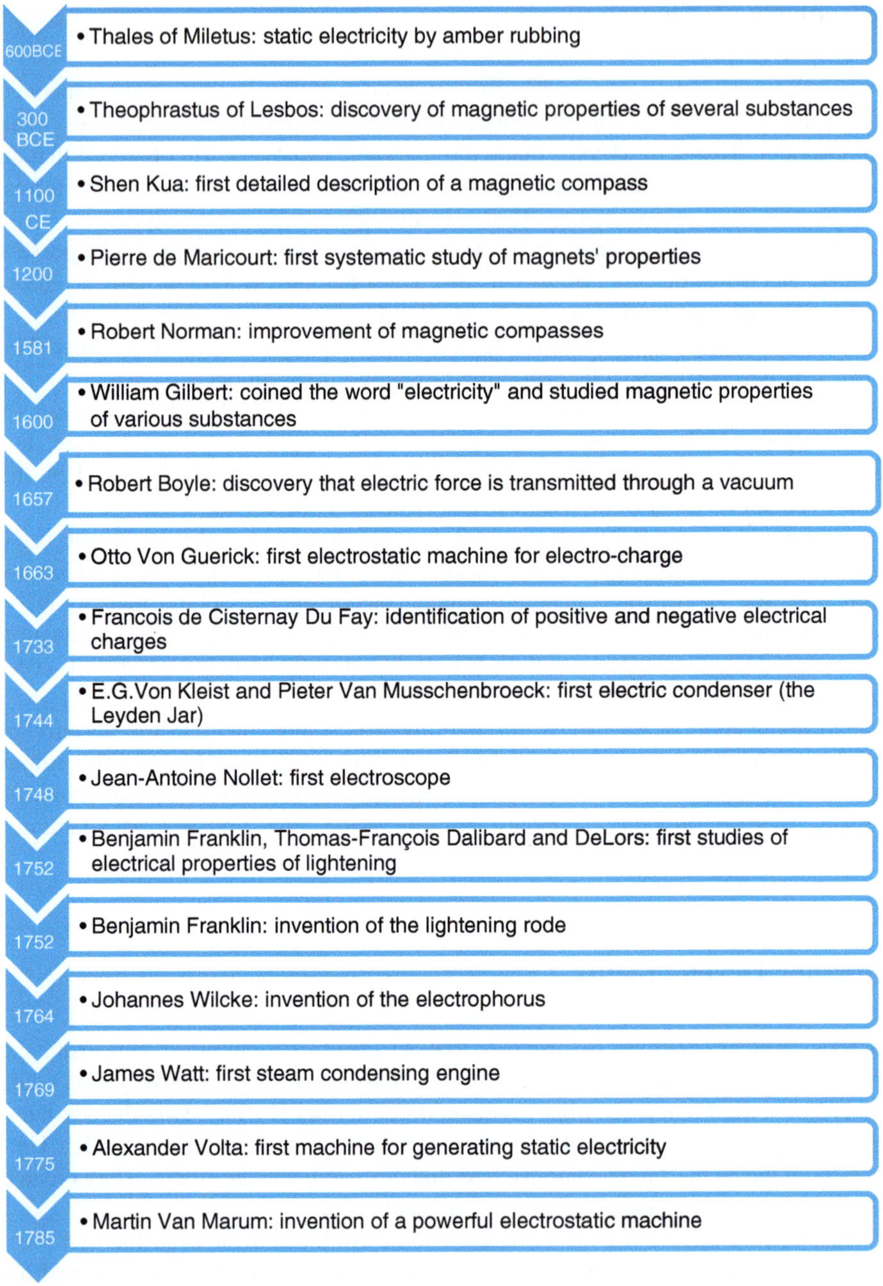

Fig. 7.2 Timeline of major inventions and discoveries related to electricity and artificial lighting. *Note* Assembled from NHMFL (2011) and NYT (2011)

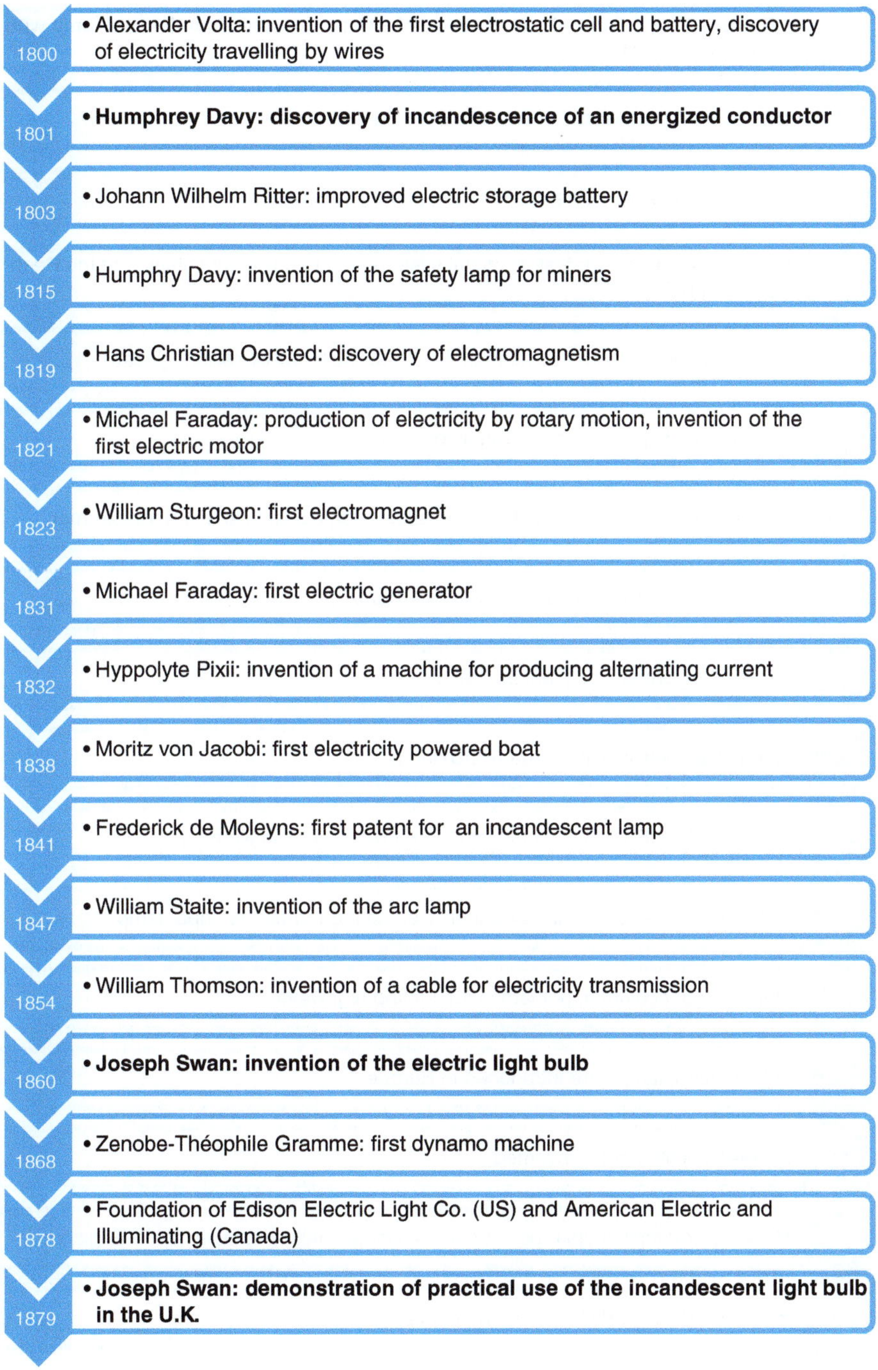

Fig. 7.2 (continued)

1879	• **Thomas Alva Edison: demonstration of a deep vacuum incandescent lamp with carbon filament**
1881	• First dynamo-powered turbine in Niagra Falls
1882	• First hydroelectric power station in Appleton, Wisconsin
1882	• Thomas Alva Edison: first electrical system for incandescent lightning at the Pearl Street in NY
1883	• Thomas Alva Edison: Invention of the "three-wire" transmission system
1885	• Niagara Falls power station begins operation
1886	• William Stanley: Invention of the first alternating current (AC) and transformer
1888	• Nikola Tesla: Invention of the induction motor
1889	• First AC hydroelectric plant in Oregon
1901	• First High Density Discharge (HDD) lamp
1904	• Thomas Alva Edison: incandescent lamp with tungsten filament
1927	• Friedrich Meyer and Hans Spanner: the first high-pressure vapor lamp
1933	• First use of sodium vapor lamps for lighting highways
1934	• Fluorescent lamps are introduced in Europe
1934	• Coiled-coil filament is invented
1956	• First commercial nuclear power station, Calder Hall, is established in England.
1959	• Invention of the dichroic (reflector) lamp

Fig. 7.2 (continued)

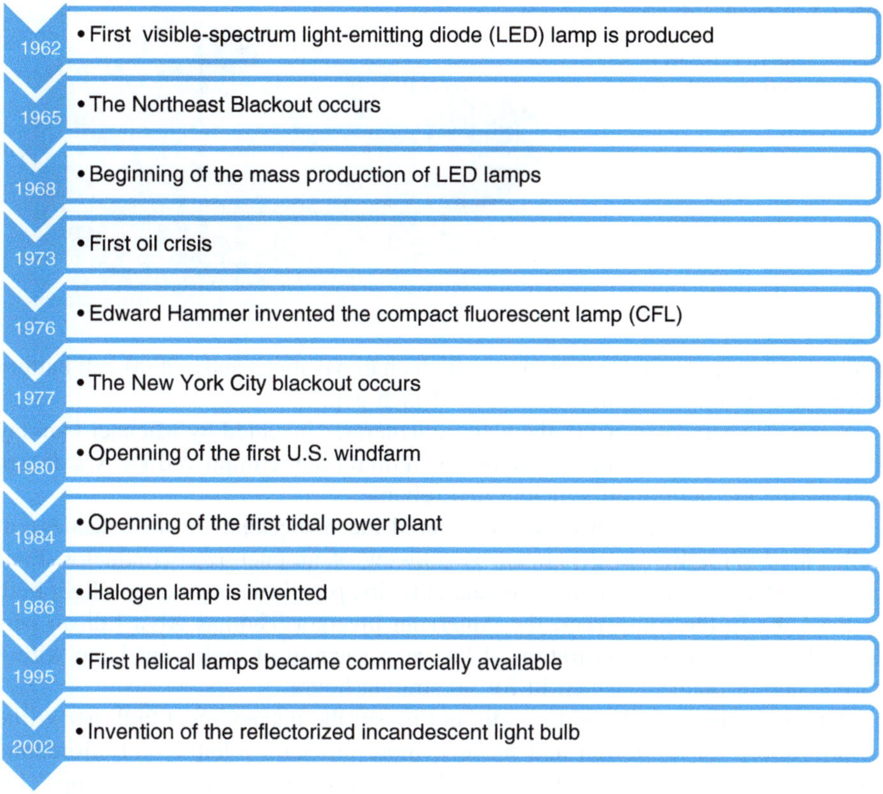

Fig. 7.2 (continued)

It appears, however, that electrical properties of lightening had been known to the Chinese for centuries. Thus, the ancient Chinese character 電 [*diàn*] stands for both electricity and lightening and means a struck of lightning from rain clouds, 雨 to the crop field, 田.

Two other major breakthroughs in electricity research were achieved in 1769, when James Watt invented the first steam condensing engine, and in 1775, when Alexander Volta developed the first machine for generating static electricity.

Twenty-five years later, Alexander Volta made another ground-breaking discovery, demonstrating that electricity can travel between places along a metal wire.

Discovery of Artificial Illumination

The first known attempt to use the newly discovered electricity force for artificial illumination was made only in 1801, when the British chemist and inventor Sir Humphrey Davy discovered the incandescence of an energized conductor. Later

Fig. 7.3 A burning Roman period oil lamp, second century C.E. (from the B.A. Portnov collection)

Davy developed the first arc lamp, in which charcoal sticks and a 2000-cell battery were used to create a light arc across a 4-inch gap.

Some 30 years later, using the electric generator invented by Michael Faraday in 1831 and known as Faraday's dynamo, William Staite improved Davy's design making the arc lamp more efficient and reliable.

In 1841, Frederick de Moleyns registered the first patent for an incandescent bulb, in which light was generated by glowing of an energized conductor. Four decades later, Joseph Swan demonstrated the first practical incandescent light bulb in the U.K. In the same year, the American inventor Thomas Alva Edison performed his famous demonstration of his deep vacuum incandescent lamp with a carbon cotton filament, which lit for 40 straight hours.

Although Thomas Alva Edison did not invent the incandescent bulb *per se*, he greatly improved its design and made it more efficient, reliable and affordable. Thus, according to Edison's 1880 U.S. Patent No 223,898 "Electric Lamp," his intention was designed "to produce electrical lamps giving light by incandescence, which lamps shall have high resistance, so as to allow of the practical subdivision of the electric light (Fig. 7.4)."

As Thomas Alva Edison promised in his 1879 interview during the first public demonstration of his incandescent lamp, "*We will make electricity so cheap that only the rich will burn candles.*" He made good on his promise. Several years later, Edison developed and installed the first electrical system for incandescent lightning at the Pearl Street in NYC, thus demonstrating, for the first time, that electric lighting can be used on a wide scale for illumination of public spaces.

As a major step towards achieving his goal, in 1883, Thomas Alva Edison developed the "three-wire" electricity transmission system similar to that used today.

Introduction of Artificial Lighting on an Industrial Scale

Two years later, in 1885, the first major Niagara Falls power station began operation followed by other major power stations in the USA and Europe. In 1904, Thomas Alva Edison developed the first incandescent lamp with a tungsten filament, making the electric lamp design much more reliable and efficient.

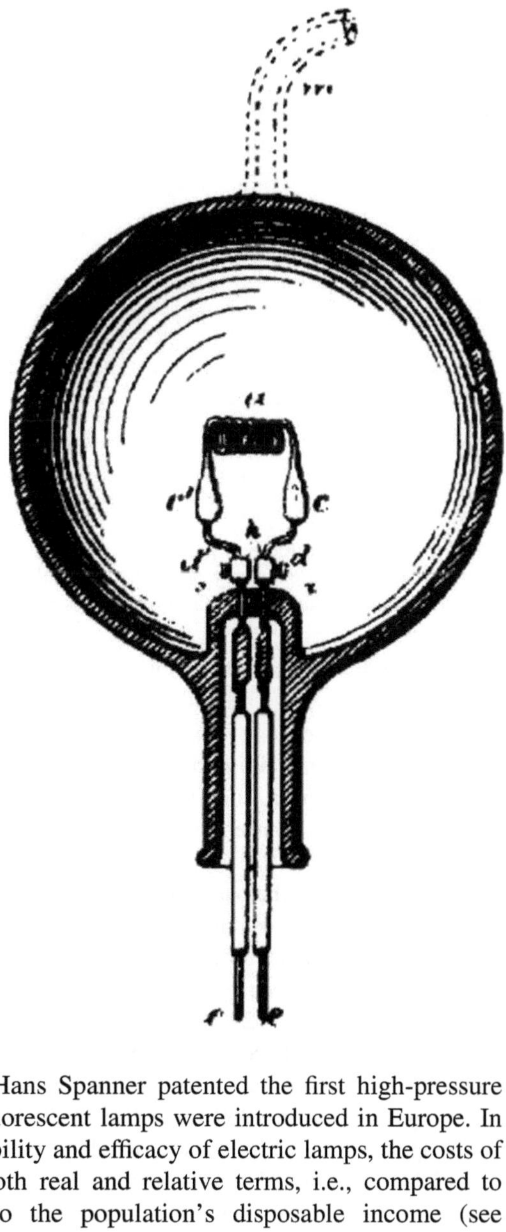

Fig. 7.4 Incandescence electric lamp (T.A. Edison's 1880 U.S. Patent No 223,898, "Electric Lamp")

In 1927, Friedrich Meyer and Hans Spanner patented the first high-pressure vapor lamp and already in 1934 fluorescent lamps were introduced in Europe. In line with improvements in the durability and efficacy of electric lamps, the costs of electricity had also declined, in both real and relative terms, i.e., compared to present-day prices and relative to the population's disposable income (see Fig. 7.5).

If in 1886, 1 kWh of electricity was worth US$ 1.32 (in 2000 prices), 120 years after, in 2007, its average price for consumers in the USA dropped to less than 8 cents per 1 kWh, that is nearly 20-fold.

This price drop was attributed to two main trends. First, electricity production facilities have become larger and more efficient, and, second, transportation costs

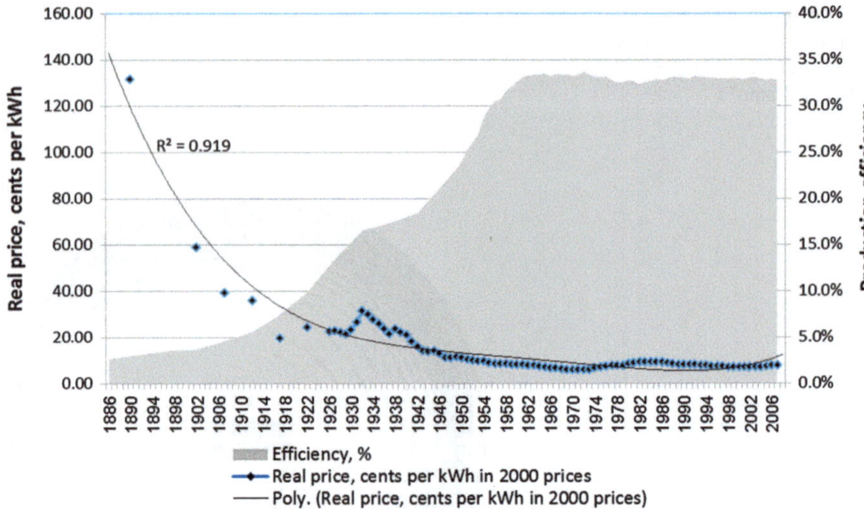

Fig. 7.5 Changes in the real prices of electricity since the 1880s (US cents per kWh in 2000 prices; *left* axis) and efficiency of electricity production from coal (*right* axis). *Note* Calculated and diagrammed using the Santa Fe Institute's electricity database (McNerney et al. 2011)

of fuels used for electricity production have been constantly dropping, at least until recently.

Although the costs of fossil fuels has increased in recent decades, this process was offset, at least in part, by introduction of energy alternative sources for electricity production, such as nuclear energy and other renewable energy sources, such as wind and solar energy. The use of these alternative energy sources made it possible to keep the costs of electricity at their present-day relatively low levels.

In parallel, over the past decades, population incomes increased in real terms, in many countries across the globe, making electricity for lighting even more affordable. Thus, according to the US Census Bureau, only during the past 40 years, real *per capita* income in the USA increased nearly threefold: from $8,980 in 1959 to $21,587 in 1999 (in 1999 prices). According to the US Economic Research Service (USERS) International Macroeconomic Database, real *per capita* incomes in the rest of the world also increased, albeit at a slower rate: from $3,138 in 1969 to $5,675 in 2010.

Dropping prices (relatively to disposable incomes) made electricity more affordable. Combined with considerable improvements in electricity infrastructures, electrification rates have increased rapidly, reaching impressive levels in both developed and developing countries alike, especially in urban areas. As Table 7.1 shows, in the world as a whole, electrification rates worldwide reached nearly 79 % and nearly 94 % in urban areas, where most population resides today.

In 1920, advocating his "State Plan for Electrification of Russia" (GOELRO), Vladimir Uliyanov-Lenin, the supreme leader of the Russian Bolsheviks, stated: "Communism is Soviet power plus the electrification of the whole country."

Table 7.1 Electricity access today—regional aggregates

Region	Population without electricity (millions)	Electrification rate (%)	Urban electrification rate (%)	Rural electrification rate (%)
Africa	587	41.9	68.9	25.0
North Africa	2	99.0	99.6	98.4
Sub-Saharan Africa	585	30.5	59.9	14.3
Developing Asia	799	78.1	93.9	68.8
China and East Asia	186	90.8	96.4	86.5
South Asia	612	62.2	89.1	51.2
Latin America	31	93.4	98.8	74.0
Middle East	22	89.5	98.6	72.2
Developing countries	1,438	73.0	90.7	60.2
Transition economies and OECD	3	99.8	100.0	99.5
World	1,441	78.9	93.6	65.1

Source Compiled from the IEA World Energy Outlook 2010

Under Lenin's successors the GOELRO plan was implemented in full, but, evidently, Lenin's promise of communism has never materialized. Nevertheless, both workers in cities and peasants in remote Russian villages are no longer confined to their traditional *luchinas* (long and narrow splinters of wood burnt for lighting) and can now light up their houses at night and enjoy full advantages of nighttime activities, such as shiftwork and entertainment, wherever available.

Concluding Remarks

As a result of a rapid electricity proliferation, human population has become virtually independent of natural L/D cycles. In fact, our homes and work places are lit today to the levels comparable to day-time natural illumination (see Fig. 7.6). As a result, humans do not have dark nights *per se*, unless they choose so.

Rather few people sleep nowadays in complete darkness, and many are active well after the dusk, sleeping late in the morning and during daytime. Even low intensity light levels might be a problem as their effects can accumulate and in such a case it is completely different from moon and sun light (Haim et al. 2011). Light pollution at this time is defined mainly by astronomers and appears to be increasing (see Chap. 8). As we note elsewhere in this book, this process is mainly due to a massive increase in short wavelength illumination, emitted by modern

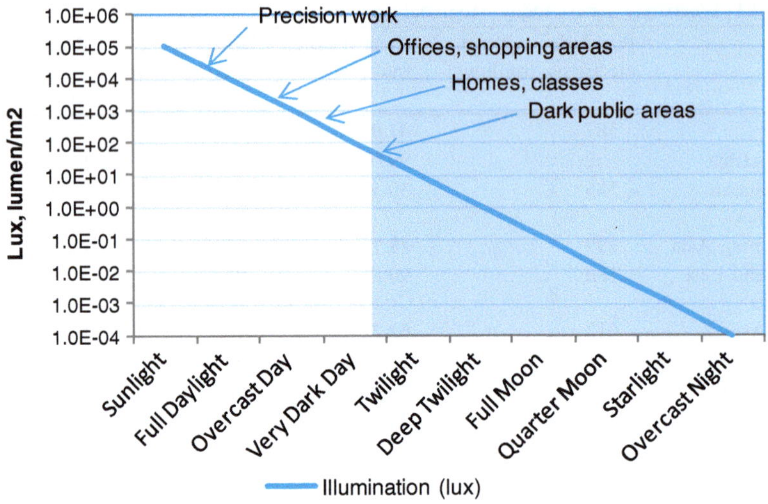

Fig. 7.6 Illumination levels in today's homes and work places compared to different natural lighting conditions (Lux). *Source* Diagrammed using data from the Engineering toolbox (http://www.engineeringtoolbox.com/)

light bulbs considered to be energy-saving and thus "environmentally friendly." A full range of implications of this change, which has happened over a relatively short period of some 120–130 years, for human evolution in general and human health in particular is yet to be determined. We shall attempt to discuss these implications in the following chapters.

Chapter 8
Biological Definition of Light Pollution

Abstract Whereas light pollution is well defined by the astronomers, no biological definition of this phenomenon has been given so far. In giving such a definition, various variables of nighttime illumination, such as intensity, wavelength, duration of exposure, frequency of exposure and timing within the dark period, should be considered. A critical point is that light pollution suppresses MLT production and disrupts daily rhythms entrained by photoperiod and thus initiates a stress response, all of which are biological consequences of exposure to LAN.

Keywords Light pollution • Light at Night (LAN) • Continuous exposure • Non-image forming photoreceptors • Bipolar cells • Melanopsin • MLT • Shift-workers

> *The shepherd drives the wolf from the sheep for which the sheep thanks the shepherd as his liberator, while the wolf denounces him for the same act as the destroyer of liberty. Plainly, the sheep and the wolf are not agreed upon a definition of liberty.*
>
> Abraham Lincoln

Light Pollution: Definition by Astronomers

Increasing urbanization worldwide, on the one hand, and more efficient transformation of electrical energy into illumination, on the other, brought about a new situation, which was first known as LAN and is now considered "light pollution" and even "light toxicity". Astronomers were the first who paid attention to this phenomenon, by pointing out that dark nights are disappearing. The international meeting that took place in April 2007 in La Palma, Canary Islands, Spain resulted in a document known today as the "La Palma Declaration." The main aim of this declaration was to draw attention to the fact that artificial illumination today interferes with our ability to observe the stars. The declaration called for the

A. Haim and B. A. Portnov, *Light Pollution as a New Risk Factor for Human Breast and Prostate Cancers*, DOI: 10.1007/978-94-007-6220-6_8,

defense of the dark night sky so as to continue to see the starlight. In the declaration, light pollution was defined as follows:

(1) Environmental pollution consisting of the excess of harmful or annoying light.
(2) Wasted light from city and outdoor lights that makes it hard to see the stars at night, and
(3) Misdirected, unshielded, excessive and/or unnecessary night lighting aimed upwards or sideways, scattering light across the atmosphere, brightening the night sky, while diminishing the view of it.

Another definition of light pollution given by astronomers is that light pollution is the introduction by humans, directly or indirectly, of artificial light into the environment (Blask et al. 2005). The astronomers thus emphasized the main direct result of light pollution from their point of view, which is the excessive illumination of night-sky by artificial light.

Definitions of Light Pollution in Biomedical and Chronobiological Terms

It took a little longer for the life and biomedical sciences to realize that LAN coming from artificial sources is environmental pollution. During the 1980s and 1990s, empirical evidence about negative effects of LAN exposure was accumulated mainly from laboratory experiments and from studies of specific population cohorts professionally exposed to LAN, such as shift workers. Therefore, during that period, light pollution, was considered mainly as a professional problem, a sort of occupational risk, specifically in flight attendants and nurses (for a detailed review of the evolution of studies on the topic, see Chap. 11).

The exposure to LAN of shift-working nurses and its negative health consequences (specifically in relationship to BC) were investigated in several cohort studies, carried out, nearly simultaneously, in the USA and in Europe. These studies revealed strong relations between shift-working and BC risks, with women working for more years in nighttime shifts being at higher risk of developing BC than those working for short periods (see Chap. 11).

As a result of these studies, a subcommittee of the World Health Organization during a meeting that took place in Leon, France in 2007 decided that shift work is a 2A risk factor for cancer in humans, that is, a "potential human carcinogen" (Straif et al. 2007). While cancer-related risks associated with shift-working (often taken place at night in brightly lit environments) were thereby acknowledged, light pollution *per se* was not mentioned in this decision as a potential cancer risk factor.

Biologists studying animal ecology using animals from different taxonomic groups published the first edited book, entitled "Ecological Consequences of Artificial Night Lighting" (Rich and Longcore 2005). Summarizing results of previous studies on the topic, they stated that artificial "lighting probably suppresses pineal MLT production and disrupts circadian rhythms in mammals."

As various effects of LAN on different animals are described in the literature (and discussed elsewhere in this book), many of these studies lead us to a conclusion that LAN is involved in suppression of pineal MLT production, one of which main roles is to transfer the light and dark signals to the cells, tissues and organs in mammals. High levels of MLT means dark phase (night) while marginal values means light phase (daytime).

Thus, according to the mammalian model, the dark signal transformation from the environment to the pineal gland contains interaction with melanopsin, the photo-protein existing in the bipolar ganglions of our retina, and discovered only about one decade ago (see Chap. 6 for more detail). The separation of the photoreceptors to image and non-image forming photoreceptors (IFP and NIFP respectively) could make chronobiologists focus on NIFP and their response. Furthermore, as the latter are sensitive to short wave length illumination, it was concluded that the vertebrate major oscillator works blue (Erren et al. 2008).

Illumination Characteristics that may be Useful for a Definition of Light Pollution

As discussed in Chap. 2, light can be characterized by several properties (such as intensity, wavelength, duration of exposure, frequency of exposure and timing), and understanding of the role of each one of them may help us to find an appropriate biological definition for light pollution. The results of Cajochen and coworkers, discussed in Chap. 6, revealed the importance of wavelength on MLT suppression, with the exposure to short wave length of 450 nm being most effective. Given to humans, such exposure suppressed MLT production, increased alertness, limited body temperature and decreased heart rate. Exposure to the same light intensity at the same time and for the same duration but at a wave length of 550 nm did not result in a similar response as MLT was not suppressed, alertness decreased, while heart rate and thermoregulatory system behaved in a similar way to the control group that was kept under dark conditions without light interference (Cajochen et al. 2005).

These results can be attributed to the fact that the bipolar cells (NIFP) contain *melanopsin* which is sensitive to short wave length illumination and controls the circadian functions of the humans. The suppression of pineal MLT production should, therefore, be included in the definition of light pollution as the results of various experiments show that short wave length illumination rather than medium or long wave length is effective.

The duration of exposure and its timing should also be considered in giving a biological definition of light pollution. As Hoffmann's (1979) study revealed, the exposure of short day (SD) acclimated Siberian hamsters (*Phodopus sungorus*) males for one or five minutes to white light in the middle of scotophase resulted in increase of testis mass and of accessory glands and values were similar to those of long day (LD) acclimated hamsters, conditions in which this species reproduces. As *P. sungorus* is a long day breeder SD-acclimation through increased levels and secretion duration of MLT suppressed the testis and accessory glands' size and function, while the suppression of MLT production through exposure to LAN removed the suppression effect and an increase in both variables was noted. Therefore, LAN can be treated as a factor that through its suppression of MLT changes seasonality (Haim et al. 2005) as shown in Fig. 8.1.

A question we need to ask is as follows: *Will the same exposure duration at the same light intensity have a similar effect on humans in regards to MLT production and secretion*? Another question is: *Will or will not such an exposure be time dependent*?

To the best of our knowledge, no experiments have been carried out on humans. The results of such experiments will further strengthen the LAN exposure and the MLT suppression nexus.

In animal models, the results of various studies were quite definite, showing that suppression of MLT by LAN increased proliferation of BC and PC cells *in vivo*. In one of such studies (Haim et al. 2010), prostate TRAMP cells inoculated to mice were used as a model for understanding such relations. Inoculated mice were exposed to short day (SD) and long day (LD) conditions. Under the LD conditions the tumor rate of growth was significantly higher than under SD-conditions. However, MLT treatment to LD mice significantly decreased rate of growth while exposure of SD-mice to 30 min of florescent white cold light at an intensity of 450 lux and a dominant wave length of 470 nm increased tumor size.

In another experiment carried out in the Laboratory of Chronobiology at the University of Haifa, SD-female mice were inoculated with 4T1 mice BC cells. After inoculation mice were divided to three groups: (1) SD-acclimated used as control; (2) SD-acclimated with light interference of 30 min seven hours after lights went off (LI-mice) and (3) LI-mice treated with MLT given in the drinking water one hour before lights went off till the end of the dark phase (LI + MLT). After 21 days a significant difference was noted in the tumor volume of the three groups where it was the largest in LI-mice and smallest in LI + MLT treated mice. In SD- mice it was intermediate larger than in LI + MLT group. The results published in scientific literature, together with ours emphasizes the major role MLT has as a hormone produced at night under dark condition. Disruption of its secretion by short wave length illumination suggests that such a type of

Fig. 8.1 The effects of LAN on seasonal acclimatization. SD = short day; LD = long day; L = light phase; D = dark phase

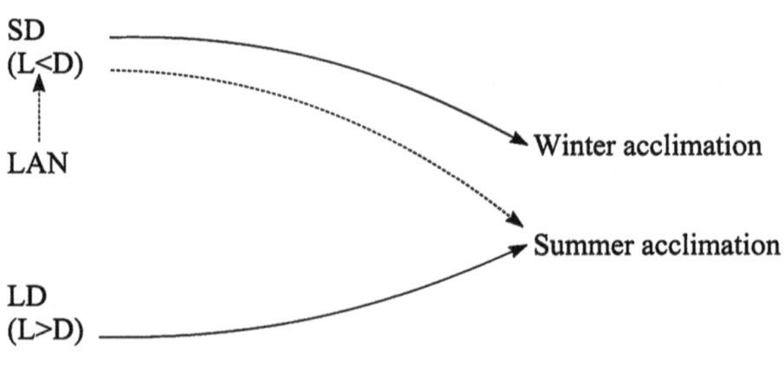

illumination is a source of pollution or of toxicity. Therefore, in any definition of light pollution MLT suppression should thus be a part of the definition.

The total duration of exposure to LAN (frequency of exposure events and the duration of exposure in each event) should also be considered in the definition of light pollution. Data to support this claim come from studies of shift workers. The results of such studies revealed that in nurses whom worked for longer periods on night duties were in a higher risk group relatively to those who worked for several years only.

Another piece of evidence comes from laboratory experiments. Thus, yet unpublished results from research carried out in the Israel Center for Interdisciplinary Research in Chronobiology (ICIRC) show that exposure of mice inoculated with BC cells to LAN results in hypomethylation of the global DNA in the BC cells. The results also suggest that MLT can increase methylation and therefore it seems to reverse the negative effect of LAN. If we knew more about such processes in humans we could have scheduled shift workers into a more sophisticated timetable, thus minimizing the negative effects of LAN.

Switching on the "cold" white light, such as florescent, even for a very short time (in order of several minutes or less) may also act as a stressor, as the results of a study carried out on the social vole *Microtus socialis* by Zubidat et al. (2007) revealed. Cold white light at an intensity of 450 lux with a dominant wave length at 470 nm for 15 m, every four hours, resulted in a significant increase of hypothalamus-adrenal medulla axis as assessed from adrenaline levels. A significant increase was also noted in the hypothalamus–pituitary–adrenal cortex axis assessed by the levels of plasma cortisol. Results of a recent study (Ashkenazy and Haim 2012) which detected an increase in HSP70 as a result of LAN also supports the idea that LAN, especially short wavelength LAN may acts as a stressor (for more studies see Chap. 9), thus suggesting that a potential stress effect of LAN should be added to the definition of light pollution.

Therefore, the tentative biological definition of light pollution we suggest is as follows:

> *Light pollution is light emission from artificial sources given in the dark phase of the 24 h cycle which wavelength and/or intensity can suppress pineal MLT production, disrupt daily rhythms and initiate a stress response.*

Chapter 9
Light-at-Night (LAN) as a General Stressor

Abstract The suppression of MLT by exposure to Light at Night (LAN) and its effects on daily rhythms are well documented in the literature. Throughout the years of research it has been noted that MLT treatment could not compensate for all the effects of LAN exposure and, therefore, we can conclude that other responses may result from the exposure to LAN. The idea that LAN is a general stressor was tested in the social vole *Microtus socialis* and an increase in adrenaline and cortisol production was detected. Recently it was also shown that LAN increases the expression of the gene hsp70 and the protein HSP70 in the brain and liver of golden spiny mice *Acomys russatus*, with the response levels in both tissues decreasing with the duration of acclimation. However, in the same animals, the response of the cardiocytes increased with the duration of acclimation thus indicating that different tissues respond to LAN in different ways. The fact that MLT treatment attenuates the response to LAN points to the fact that MLT may reduce negative impacts of LAN as a general stressor.

Keywords Stressor · HA-axis · HPA-axis · Adrenaline · Corticosteroid · MLT · HSP70

> *Stress is a non-specific body response to environmental changes that endangers life.*
>
> Selye (1950)

LAN and Stress Response

The results of an experiment carried out on the social vole (*Microtus socialis*) revealed that food and energy intake did not differ between individuals exposed to light interference (LI) and LI treated with melatonin (MLT) (Haim et al. 2004). This result showed that LI is not only a signal that affects daily rhythms seasonality response by suppressing pineal MLT production, and, therefore, the study

A. Haim and B. A. Portnov, *Light Pollution as a New Risk Factor for Human Breast and Prostate Cancers*, DOI: 10.1007/978-94-007-6220-6_9,

raised the following question: *Are other systems, apart from the pineal gland and MLT suppression, affected by LI*?

One possibility could be that LI is a stressor and it can activate stress response by increasing the activity of the hypothalamus–adrenal medulla (HA) axis or hypothalamus–pituitary–adrenal cortex (HPA) axis or both of them. In a follow up study, short day (8L:16D) acclimated *M. socialis*, were exposed every four hours to 15 min of florescent light (with a dominant wave length of 470 nm) at an intensity of 450lux. Urine samples were collected in order to assess daily rhythms of adrenaline (using the analysis of its urine metabolite) and HA activity while plasma samples from blood collected at 14:00 h were analyzed for cortisol levels in SD-voles and SD + LI voles.

The results reported by Zubidat et al. (2007) and Zubidat et al. (2011) showed that activation of both HA and HPA resulted from exposure to LI thus supporting the idea that LI is a stressor and as such it may have an impact on the survival of *M. socialis* under natural conditions in the enclosures, where the voles were exposed to cold nights in winter. Assuming that produced energy had to be allocated to compensate for adjusting the homeostasis heat production for thermogenesis could not increase to the levels required for compensating heat loss.

An early field study using enclosures, in one which voles were exposed to LI while another was used as a control, it was noted that under LI of 15 min given once every hour, once every 2 h and once every 4 h, voles did not survive winter conditions not only because of being exposed to "seasons out of time" (Haim et al. 2005) but also because of exposure to a stressor.

Selye (1950) defined stress as "a nonspecific body response to environmental changes which endanger life". Individuals may elicit the appropriate response bringing themselves back to the state of homeostasis. More recent literature suggests that stress response is activated by sudden unanticipated and unpredicted events (Koolhaas et al. 2011). A question that can be asked in regards to stressors and stress response is: *Will repeated exposure to a stressor result in a decrease of response or, in other words, can a subject be acclimated to stressful conditions*?

Stress Response to LAN on a Cellular Level

On the cellular level the impact of light pulses was assessed by the expression of transcripts in the mouse brain using microarray technology in more than 200 different transcripts (Ben Shlomo and Kyriacou 2010). In another study, a light pulse given to the chicken pineal at the end of the subjective night caused, among other, the activation of genes responsive to heat shock stress in the *pinealocytes*. Expression of regulatory transcription factors, heat shock factors one and two hsf 1 and 2 (Hatori et al. 2011). The results of physiological studies on the one hand and those of cellular studies on the other showed that LI causes changes which can be considered as response to a stressor.

Are there other cellular mechanisms which can express stress respond to LI?

A cellular stress response is activation of heat shock proteins (HSP) so far an activation of HSP according to Kregel (2002) was noted as a response to various stressors such as thermal (hyperthermia), acidosis, energy depletion and an increase in free radicals. Therefore, it seemed logic to test the following hypothesis: *If LI is a stressor, exposure to LI should result in increased levels of HSP.*

This hypothesis was tested in golden spiny mice *Acomys russatus*, exposed to acute, medium and chronic periods of LI. Short day (SD, 8L:16D, lights on between 08:00 and 16:00 h). The acclimated mice were kept in a climatic chamber where during day time, cool florescent light at an intensity of 450lux and a dominant wavelength at 470 nm while a dim light of 25lux at a wavelength of 679 nm was kept constantly. For LI, day light was on for 30 min 6 h after darkness onset. Four groups of mice were sampled by removing organs as brain, liver and heart. The groups included: (1) control–SD, (2) SD + LI for two nights (acute), (3) SD + LI for seven nights (medium) and (4) SD + LI for 21 nights chronic. Levels of protein HSP70 and expression of *hsp70* were measured. The results showed a significant protein increase in both brain and liver tissues after an acute exposure followed by a decrease in the duration of medium and chronic exposures.

The gene expression increased significantly in the brain tissue of mice exposed to acute LI and decreased in medium and chronically acclimated mice. However, in the liver tissue no gene expression was noted in any of the tested groups (Ashkenazi and Haim 2012). These results indicate that LI is a stressor that activates the production of HSP70, but these results also suggest that the brain tissue can be acclimated to such interference and the response decreases with the duration of acclimation. In a follow up study we measured HSP70 and the *hsp70* gene expression in cardiocytes and interestingly the results showed an increase in protein levels and gene expression under chronic exposure relatively to acute exposure (Ashekenazi et al. unpublished data). This result raises the question in regards to the stress response of different tissues and in future studies protein levels and gene expression levels should be measured in more tissues. Furthermore, results of exposure of pregnant mothers and after birth of the young to LI may be of great importance to understand the possible damages caused in different stages of development (ontogenesis). The following question can thus be asked: *Does LI have an impact on longevity and on general health?*

Further research of these topics is of great importance and should thus be supported by national and international funds.

Oxidative stress in general increases reactive oxygen species (ROS). ROS have a negative impact on the cellular level as they alter protein and DNA molecules (Cerutti et al. 1994), and therefore play a significant role in disease and aging (Kregel and Zhang 2007). In response to increased oxidative stress organism will increase production of anti-oxidant compounds as enzymes including superoxide dismutase (SOD), glutathione peroxidase (Gpx) and catalase (CAT); the non-enzymatic anti-oxidant defenders include MLT among other molecules and is known as an effective anti-oxidant (Matés et al. 1999; Reiter et al. 2003; Garcia et al. 2010) and in stressed algae it was shown to be the first line defender in response to stress effect

(Tal et al. 2011). Results from *A. russatus* exposed to LAN increased Gpx and SDO activity thus supporting our hypothesis that LAN is a general stressor. However, MLT attenuated the response (Ashkenazi and Haim, in prep.)

The association between stress and the immune system is well established reflecting the fact that energetic resources are directed to reestablish homeostasis which is changed as a response to a stressor. In the case of limited resources energy supply may emerge from reduced energy supply to the immune system thus affecting its normal function. Future experiments in regards to LAN and our understanding of its negative impact on our health should include such topics as our knowledge on these topics is quite limited. Bearing in mind that exposure to LAN of high illumination intensity can occur in a very short time (magnitude of parts of a second) and therefore in many ways it is similar in nature to the lightening which can be considered as a stressor.

Changes from *scotophase* to *photophase*, under natural conditions are gradual in respect to light intensity. Today the most common way for controlling illumination is using on/off switchers. Instead of such switchers, the use of computerized dimmers, for controlling illumination operation, may reduce the stress effect of LAN. Further studies should thus be carried out on this topic because it should be realized that LAN is not only a source of pollution suppressing MLT production but it also affects our daily rhythms, which should jointly be considered in the definition of light pollution.

Chapter 10
Effects of Light Pollution on Animal Daily Rhythms and Seasonality: Ecological Consequences

Abstract The ongoing increase in light pollution and, mainly, that of short wavelength light emission, results not only in the human pineal MLT suppression, but also affects that of animals in their "natural" ecosystems. It also affects animals' daily rhythms, seasonality and behavioral patterns such as foraging (aerial, aquatic, and terrestrial) due to changes in the distribution patterns of their prey. The disruption of navigation by animals, such as young sea turtles' navigation into sea-water, is also affected by urban illumination and sky glowing around cities at night. As LAN has an effect on the entrainment of our biological clock, by disrupting daily rhythms, it results in mal-functioning of the immune and physiological systems. Because these features are common to humans and animals, their similarity makes it possible to use animals for testing and developing models for sustainable nighttime illumination.

Keywords Prey · Predator · Aggregation · Foraging · Aerial · Plankton · Behavioral patterns · Navigation · Animal models · Sustainable illumination

> *Our diurnal bias has allowed us to ignore the obvious, that the world is different at night and that natural patterns of darkness are as important as the light of day to the functioning of ecosystem.*
>
> Rich and Longcore (2005)

Effects of LAN on Ecosystems

The fact that LAN, as a source of environmental pollution, may have a negative impact on natural ecosystems is noted by Longcore and Rich. In their comprehensive review, entitled "Ecological Light Pollution" published in *Frontiers in Ecology and the Environment* in 2004, the authors drew our attention to the following facts about such pollution: (1) LAN includes increased illumination that

A. Haim and B. A. Portnov, *Light Pollution as a New Risk Factor for Human Breast and Prostate Cancers*, DOI: 10.1007/978-94-007-6220-6_10,

can be chronic or periodical, causing unpredicted changes in illumination and direct glare. (2) Animals can be disorientated by additional illumination as we know from the case of young sea turtles, hutching from their eggs, and navigating their way to the sea; apart from navigation, night illumination may affect foraging, reproduction and communication. (3) LAN has an impact on inter-relations between organisms which may have significant implications on the ecology of communities.

Although our book deals mainly with human health models in regards to BC&PC, understanding what happens to other species as a result of LAN exposure, may help to understand the full extent of LAN effect on humans.

In this chapter we deal with negative impacts of LAN on the ecology of organisms in their natural ecosystems. As well known, the effects of LAN on the entrainment of biological clock and its effects on daily rhythms and seasonality are common to humans and diurnal mammals that show daily rhythms similar to those of humans, where the acrophase of motor activity and body temperature occur during the photophase while that of pineal MLT secretion occurs during the scotophase.

LAN and Interrelations Between Predators and Prey

In their review entitled the "Dark Side of Light at Night", Navara and Nelson (2007) deal with the ecological consequences of the LAN phenomena. One of the studies cited in this review was carried out in the open desert terrain and reported by Clarke (1983). As this study showed, during bright moon nights, the deer mice *Peomyscus maniculatus*, and the short-eared owls, *Asio flammeus*, behave differently. While the former, the prey, decreases its activity on full moon nights, the owl, the predator, increases its activity. The behavioral changes indicate that animals from different trophic levels behave differently upon exposure to the very same environmental conditions. Full moonlight is a natural event which happens every lunar month. Therefore, these behavioral differences are a natural response which has been presumably developed under natural selection and can be considered as adaptation.

However, we must keep in mind that under full moonlight, illumination intensity is much lower than sky brightness emerging from city-sky glowing (Cinzano et al. 2001). Another difference is that moonlight does not contain short wavelength illumination, a type of illumination which is often dominant in cities' glow. It is also important that full moon nights are anticipated by the animals and occur for few nights every month, while LAN is constant and even increase in intensity over time. Therefore, the effects of LAN on the predator-prey relationship are likely to be more pronounced than the effects of moonlight.

Insectivore Bats and Changes in Prey Distribution

In regards to foraging, a mammalian group strongly affected by LAN is the insectivorous bats (or *Microchiroptera*). The short wavelength illumination which is used in many public places attracts insects which aggregate close to sources of such illumination. Some bat species follow these aggregation sites and they prey on insects (Rydell 2005). Beyond doubt the change in the spatial distribution of the prey has an effect on the bat species with specialized sonar apparatus, which have an advantage when foraging in open areas under dark conditions, but becomes less advantageous when preying on insects concentrated close to blue light sources. As a result, we may end up by losing insectivore bat species, which have an important role in pest control and great efforts are being made in regards to their conservation.

Aquatic Ecosystems: Plankton Migration and LAN

Aquatic pelagic organisms in fresh and sea water are sensitive to light, where zooplankton demonstrates vertical migration in response to light intensity. Under high light intensities, the zooplankton sinks in the direction of the benthos. However, when light intensity is low, it moves in the direction of the upper pelagic. This is a daily movement initiated by the differences between light intensities penetrating the water. Under natural conditions the highest light intensities during nighttime are close to those of a full moon (Haney 1993; Gal et al. 1999).

The daily plankton movement is important for fish foraging. Within the zooplankton we can distinguish between grazing species which are smaller, relatively to the carnivore plankton species, which are larger and avoid exposure to light. This is a natural response to avoid predation by fish. Therefore, even under a dim light zooplankton will not show a vertical migration in order to avoid predation by fish, but at the same time they are depriving themselves from foraging. Fish show a vertical daily migration. For instance, in Patagonia, Rechencq et al. (2011) showed that fish use the lower darker part of the lake at day time when they are inactive while in the dark phase they migrate vertically for foraging.

As Moore et al. (2001) noted, in most lakes located near cities, zooplankton of different systematical groups do not exhibit a vertical migration, which can be explained by exposure to LAN that suppresses migration towards upper layers, where water is warmer and has more food compared to lower layers. The removal of LAN is therefore accompanied by vertical migration.

Does zooplankton responds to night with full moon light in the same way it does to LAN? The answer to this question is no because full moon nights occur only for a few nights while exposure to LAN is every night. In the Lake of Galilee, for instance, a decrease in fish populations has been noted in the last 20 years. Among the commercial fish which population decreased significantly, is the bleak fish, *Acanthobrama terraesanctae*. These fish are known to forage on zooplankton at night and breeds under short days. As a result, LAN has a negative impact on both foraging and reproduction of this fish species. Because the lake is a touristic attraction, it is illuminated by high intensity illumination which could be a reason for the reduction in the fish population. However, this phenomenon should be studied further in order to understand the mechanism by which LAN may affect fish reproduction.

In regards to four different aquatic species, the implications of LAN were studied by Bruning et al. (2011). The authors of this study concluded that LAN has an effect on hatching and initial swim bladder filling by masking the day–night-change. The study also revealed that the reactions were specific for different species and the increase in variation indicated a lack of diurnal triggering, although harmful effects of LAN were not identified on early life stages.

The Black Bird and Pigeons' Responses to LAN

The black bird, *Turdus merula*, is a common species in the Mediterranean ecosystems of Israel. Its populations are seen all around the year and they breed under increasing photoperiod. These birds live in our gardens, waking people up in the morning when the male birds start singing. In past years, many black birds became nocturnal, as they sing and forage under short or long days at the first part of the dark phase and even at midnight or later. Several questions can be asked in regards to the described phenomenon:

Are diurnal birds active at night time more vulnerable in regards to predation? Why is it so common in the black bird to extend its activity into the dark phase of the 24 *h cycle under artificial illumination and not for other song birds as the Jay for instance that did not change its daily activity patterns? Do the black birds show more elasticity in regards to daily rhythms? As now black birds sing and forage at midnight which means they changed their daily patterns of activity, or changed their daily rhythms, is it an advantage? Although, this change may be considered a great success for the black birds, is it really a success? May the change in their temporal organization lead to breeding out of season? Are they (not only the adults but also the young) exposed to predators they cannot cope with? Can this new phenomenon be considered as an adaptation?*

It is difficult to tell for sure what will happen to this species (as to other species who shifted activity in a similar way) but they are definitely facing an environmental challenge which is an outcome of human activity. We do believe that ornithologists studying bird behavior will try to understand the consequences of these temporal modifications.

There is another interesting example of behavioral change in birds, presumably attributed to increased light pollution. Cities in Europe and around the Mediterranean Basin are inhabited by pigeons and, in Scandinavian cities, in the northern latitudes, their populations have been moving to the north in the past decades without migrating in winter. As food is not a limiting factor and photoperiod is longer due to LAN, pigeons can exist and reproduce throughout the year, thus becoming pests with increasing damages to buildings in urban areas (Haim et al. 1979).

New Food-Webs as a Result of LAN

As many insect species are attracted by LAN (mainly blue short wavelength light) they aggregate around sources of such illumination, many of which are connected to walls, and places that insectivore bats cannot reach. The aggregations of these insects near light sources attract geckoes who prey on them. If illumination releases much heat (such as e.g., incandesced bulbs), ectothermic geckoes efficiently prey on the insects. The third animal that joins such a food chain is the cat that preys on the geckoes, thus creating artificially a new food chain nonexistent in nature. *We can ask how sustainable are these human made food-webs*?

In order to establish sustainable illumination we need to look for technologies which we can use for producing light that will be "environmental friendly" in the sense that it will have minimal negative effects on the environment including both humans and animals. Our efforts to increase the photophase have had a negative impact on our surrounding but also economic implications, such as in the case of decreased fish crop. Therefore, ignoring LAN as a source of pollution on the natural ecosystems may not be a clever step, and, therefore, sustainable illumination policies should be sought and implemented to provide reasonable levels of nighttime illumination required by human activities, but with minimal damages to the natural ecosystems.

Chapter 11
Light Pollution and Hormone-Dependent Cancers: A Summary of Accumulated Empirical Evidence

Abstract A possibility that human body may be affected by ambient light was raised, apparently for the first time, by the Israeli physician Dr. Phillip Cohen in 1970. Since then the circadian rhythm theory (CRT) has received considerable support, from both animal models and human studies. As we know today, the link between artificial light and its potential health effects may be attributed to two interdependent mechanisms—inhibition of melatonin (MLT) secretion from the pineal gland by direct exposure of human vision system to LAN and daily rhythm disruption. Although these effects are discussed in detail in separate chapters of this book, in this chapter, we review, in brief, supporting empirical evidence accumulated to date.

Keywords Light pollution · Daily Rhythm Disruption · MLT · Light At Night (Lan) · Hormone Dependent Cancers · Empirical Studies · High-Low Risk Groups · Light–Dark (L–D) Cycle · General Surveys · Specific Group Studies · Population Level Studies · Laboratory Experiments · Accumulated Evidence

> *Daylight changes, which influenced the organism during the millions of years of evolution, may have had its reflection in an endogenous year rhythm, the environmental daylight changes working as a time-keeper only*
>
> Phillip Cohen (1970)

Introduction

As we previously discussed elsewhere in this book, the link between LAN and BC&PC may be explained by two interdependent mechanisms. Although it may be difficult to separate them, for simplicity's sake, we should term them the "LAN-MLT path" and "LAN-Daily Rhythm Disruption" (LAN-DRD) path. The former mechanism is well researched and better understood. It is discussed elsewhere in

A. Haim and B. A. Portnov, *Light Pollution as a New Risk Factor for Human Breast and Prostate Cancers*, DOI: 10.1007/978-94-007-6220-6_11,

this book and can be summarized as follows. Artificial light which reaches human eye retina during nighttime reduces the nocturnal production of the pineal melatonin (MLT) hormone. Blue short-wave illumination, commonly used today for both indoor and outdoor lighting, is most effective in MLT suppression. Impaired secretion of MLT (a hormone with tumor inhibiting properties) may result in estrogen receptor affinity and thus increase the susceptibility of humans to BC&PC.

If the "LAN-MLT" hypothesis is correct, then any person exposed to high intensity of artificial light during nighttime (when MLT is chiefly produced) is likely to be in a high risk group for BC&PC. First and foremost, such population cohorts include nighttime shift workers (that is, hospital physicians, nurses, bus drivers, undertakers, cashiers and shop assistants, flight attendants, etc.). This "high risk" group may also include any person who works, studies, or regularly attends places of entertainment late at night, as well as people suffering from insomnia or sleep disturbance, and/or persons sleeping with full lights on during nighttime.

By the same logic, the group of people in a relatively low BC&PC risk group, according to the LAN-BC&PC hypothesis, includes the totally or partially blind; residents of extreme latitudes with long "polar" nights; long sleepers; people with high sensitivity to light and thus spending most of their time in the dusk; people working in dark environments (e.g., miners, infrastructure maintenance workers, etc.), and residents of remote rural areas, especially in low resource countries with underdeveloped electricity networks (see Fig. 11.1).

The second mechanism of LAN influence (that is, the LAN-DRD path) is less well researched and understood. According to its simplified interpretation, even if a person is totally blind or keeps eyes closed at all times, but is active at night and sleeps during daytime, his or her circadian gene clock function, formed during the millions of years of evolution, may be disrupted. This would stress the immune function and thus increase the susceptibility of such a person to various diseases, including BC&PC.

If this hypothesis is correct, then any person, who regularly works, drive a car, ride a bus or regularly attends places of entertainment late at night and sleeps during daytime, is at high risk of developing BC&PC, no matter whether such a person is directly exposed to high intensity LAN or not.

As we previously noted, the above mechanisms of LAN influence are not independent though. Indeed, any person who works or attends places of entertainment at night *must be exposed to LAN*, unless he or she is doing so in complete darkness which is highly unlikely. Likewise, a person who is exposed to LAN, either from indoor or outdoor sources, might have trouble sleeping, and that person's normal daily cycles would be disrupted. As a result, the above mechanisms are likely be interrelated, enhancing and sustaining each other, with LAN being necessary for activation of the other (that is, of nighttime activities) and *vice versa*. In this dichotomy, MLT may potentially act as a mediator, helping to translate photoperiodic information into immune responses, as studies reviewed below suggest.

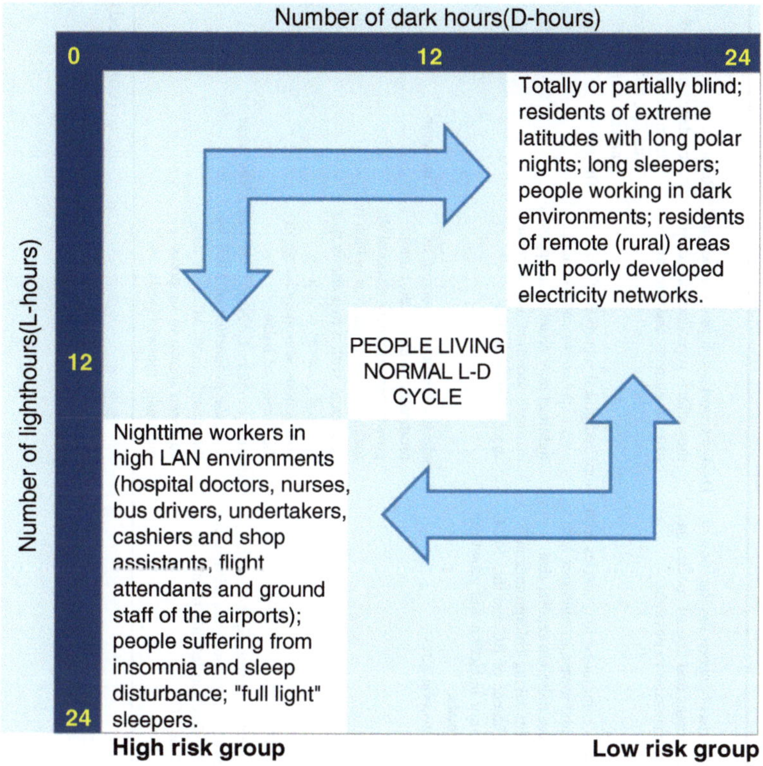

Fig. 11.1 "High-low" risk population groups under the LAN-MLT hypothesis

Summary of Accumulated Empirical Evidence

Although we discuss each of the above mechanisms of LAN-BC&PC association in some detail in separate chapters of this book, in this chapter, we shall summarize the most important, in our view, studies which have led to the development of the present day "state-of-the-art" knowledge about the LAN- BC&PC association (see Table 11.1). In doing so, we shall divide our discussion onto four separate topics: (a) general surveys; (b) studies of specific population cohorts, (c) population level studies, and, finally, (d) animal and laboratory tests.

General Surveys

The main credit for proposing the first scientific hypothesis which has subsequently led to the development of the modern LAN-BC&PC association theory, should apparently be given to Dr. Phillip Cohen from Kiryat Tivon, a small town

Table 11.1 Summary of main studies of the "LAN-BC&PC" association

Study	Research scope or contingent	Methodology	Main findings	Suggested mechanism of LAN influence	Limitations of the study
A. General analyses based on literature surveys					
Cohen (1970)	Literature survey and analysis	Critical analysis of accumulated empirical evidence.	Strong correlation exists between the duration of pregnancy in humans and animal species and the duration of exposure to daylight.	Daylight works as a time-keeper and thus affects reproduction activity of both humans and animals.	Reliance on indirect evidence obtained from different studies, both human and animal models.
Cohen et al. (1978)	The same	The same	Pineal calcification is commonest in countries with high rates of BC and lowest in countries with a low BC incidence; there are reports that psychiatric patients taking chlorpromazine have a lower incidence of BC and that MLT may influence tumor induction and growth in experimental animals.	Impaired pineal secretion of MLT (by e.g., exposure to environmental lighting) may result in unopposed estrogen secretion and increased susceptibility to BC.	The same
Cohen (1983)	The same	The same	Seasonality in BC detection rates.	Female sex hormones function as promoters of enhanced tumor growth are primarily responsible for the seasonal patterns, which is consistent with a hypothesis that seasonal patterns reflect an endogenous seasonal change in the hormonal balance.	The same
Stevens et al. (1992)	The same	Summarizing results of previous studies focusing on known risk factors and etiology of BC.	Two products of electric power, light-at-night (LAN) and electromagnetic fields (EMF), can suppress pineal function leading to an increased risk of BC.	Exposure to either LAN or EMF can decrease production of MLT by the pineal gland; MLT, in turn, has been shown to suppress mammary tumorigenesis.	The same

(continued)

Table 11.1 (continued)

Study	Research scope or contingent	Methodology	Main findings	Suggested mechanism of LAN influence	Limitations of the study
Stevens and Davis (1996)	The same	The same	The same	Electric power produces light at night (LAN) and electric and magnetic fields (EMF), either or both of which may alter pineal function and its primary hormone MLT, thereby, potentially increasing the risk of BC.	The same
Stevens and Rea (2001)	The same	The same	Electric light is different in its spectral properties from natural light and may contribute strongly to circadian disruption and thus become a major cause of endocrine disruption, thus contributing to a higher risk of BC in industrialized societies.	The adverse effect of electrical light on a 24-h circadian rhythm of MLT release, as well as on other physiological rhythms including the 12 h–L[light]-12 h–D[dark] sleep/wake cycle.	The same
Nelson (2004)	The same	The same	MLT synthesis and secretion occurs exclusively at night and is inhibited by direct exposure to light. The role of MLT as an immune-modulator is well established for many species, including humans; stress can impair immune function and thus increase disease susceptibility.	Photoperiodic information, mediated by MLT, influences immune responses.	The same
Pauley (2004)	The same	The same	Evidence exists that LAN exposure is linked to human breast and colorectal cancers in shift workers, while the indirect linkage of BC to LAN is further supported by laboratory experiments.	Suppression of melatonin (MLT) by exposure to light at night (LAN)	The same

(continued)

Table 11.1 (continued)

Study	Research scope or contingent	Methodology	Main findings	Suggested mechanism of LAN influence	Limitations of the study
Reiter et al. (2007)	The same	The same	There is increasing empirical evidence that LAN has the capability of elevating cancer risk. The actual mechanism is not absolutely clear: it may result from disruption of circadian rhythms and/or from a suppression of MLT due to LAN exposure, or from both, because circadian rhythms are entrained by MLT.	Negative effect of artificial light on chronodisruption and nocturnal MLT inhibition resulting in cancer initiation and growth.	The same
Stevens et al. (2007)	The same	Literature survey and critical analysis.		Circadian organization is disrupted by LAN, including the production of several hormone rhythms; there are also direct effects of LAN on neuro-endocrine systems, resulting in suppressing MLT synthesis or elevating cortisol production.	The same
B. Studies of specific population groups					
Hahn (1991)	11,769 hospital discharges recorded in the National Hospital Discharge Survey from 1979 through 1987, excluding women with diabetes.	Comparing the odds of profound binocular blindness among women with a diagnosis of BC with the odds of profound binocular blindness among women with diagnoses of coronary heart disease or stroke.	Profoundly blind women were half as likely to have BC as women who were not profoundly blind. This effect diminished substantially with age.	Suppression of MLT production by the pineal gland due to direct LAN exposure.	Potential effect of potential confounders, unaccounted by the research, such as family history, occupational exposure etc.

(continued)

Table 11.1 (continued)

Study	Research scope or contingent	Methodology	Main findings	Suggested mechanism of LAN influence	Limitations of the study
Suvanto et al. (1993)	40 Finnish female flight attendants with the average age of 30 years.	Measurements of oral temperature, alertness, and visual search performed at 2 h intervals during 2 days before the flight from Helsinki to Los Angeles, during the second and the fourth day in the USA and during the second and fourth day after the return flight to Finland.	Desynchronization of the body temperature and the phases of the alertness and visual search rhythms shifting; delaying of the acrophases of the circadian rhythms during the second and fourth day after the return flight. The acrophase of alertness was still delayed after the fourth day for 35 min and the acrophases of body temperature and visual search were 2 h 2 min and 3 h 8 min delayed, respectively.	Disruption of circadian rhythms by nighttime activities.	Small population group with no information on residential LAN exposure, and additional individual health risk factors (family history, smoking etc.). It is also unclear whether the reported elevated BC incidence is related to LAN, radio frequency exposure or circadian disruption due to shift working.
Pukkala et al. (1995)	1,577 female and 187 male cabin attendants who worked for Finnish airline companies for different time periods	Comparing standardized incidence ratios of selected cancers with the expected number of cases based on national cancer incidences obtained from the nationwide cancer registry.	Significant excess of BC (OR = 1.87 (95 % CI = 1.15 to 2.23) and bone cancer (OR = 15.10 (1.82, 54.40) was found among female workers. The risk of BC was most prominent some 15 years after recruitment.	The adverse effect of ionizing radiation during flights with estimates of the effect of reproductive risk factors only partly explaining the increased risk of BC.	The same

(continued)

Table 11.1 (continued)

Study	Research scope or contingent	Methodology	Main findings	Suggested mechanism of LAN influence	Limitations of the study
Tynes et al. (1996)	2,619 women who worked as radio and telegraph operators between 1920 and 1980.	Comparing standardized incidence ratio (SIR) of BC in the study cohort with that of the Norwegian female population as a whole.	Relative risk (RR) of BC in the study cohort was 1.5, after adjustment for age, calendar year, and fertility factors.	Exposure to electromagnetic fields may increase susceptibility to sex-hormone-related cancers by diminishing the pineal gland's production of MLT.	The same
Mawson (1998) [discussion of a previous study by Pukkala et al. (1995)]	Finnish female flight attendants with a mean of 13·9 years at work	Comparing standardized incidence ratio (SIR) of BC in the study cohort with the standard female population.	The same as in Pukkala et al. (1995)	MLT deficiency, resulting from work-associated interruptions in sleep-waking cycles (jetlag); chronic disturbances in circadian rhythms.	The same
Verkasalo et al. (1999)	10,935 women with visual impairment identified from the Finnish Register of Visual Impairment and followed up for cancer through the Finnish Cancer Registry for years 1983–1996.	Calculation of standardized incidence ratios (SIR) of BC for subgroups of women with different degrees of visual impairment.	BC risk decreased in line with a degree of visual impairment ($P < 0.04$) suggesting a dose–response relationship between visible light and BC risk.	LAN effect on nocturnal MLT secretion.	The survey was based on a relatively small population sample with different employment histories and did not include questions about residential LAN exposure; a possibility of potential recall error due to reliance on self-reported histories; no specific occupational health risks reported.

(continued)

Table 11.1 (continued)

Study	Research scope or contingent	Methodology	Main findings	Suggested mechanism of LAN influence	Limitations of the study
Hansen (2001)	7,035 (30–54yo) Danish women diagnosed with BC (of whom 497 worked at night) with individually matched controls.	Employment records obtained from pension funds' files matched with cancer registry data. BC odds ratios calculated using conditional logistic regression, adjusted for socio-economic status, age at the birth of first and last child and the number of children.	The odds ratio for BC among women who worked at night at least half of a year was 1.5 (95 % confidence interval, 1.2–1.7), with a tendency to increasing odds ratio by increasing duration of nighttime employment associated with exposure to LAN.	Adverse psychological and physiological effects associated with irregular working hours and specifically working at night.	The same
Schernhammer et al. (2001)	78,562 women participating in the U.S. Nurses' Health Study with 2,441 documented incidences of BC.	The probability of BC was linked to the duration of night shift working controlled for age, BMI, family history of BC, age of menopause, parity, etc.	A moderate increase in BC risk was observed among the women who worked on night shifts for 1–14 years (RR = 1.08 [95 % CI = 0.99, 1.18] and 15–29 years (RR = 1.08 [95 % CI = 0.90–1.30], respectively); the risk further increased among women who worked 30 or more years on the night shift (RR = 1.36; 95 % CI = 1.04–1.78).	Suppression of production of MLT with its potential oncostatic action due to LAN exposure	A study of a specific population group with additional potential risk facts (such as e.g., exposure to chemicals and a lack of control for residential LAN exposure and any additional LAN exposure outside work place.

(continued)

Table 11.1 (continued)

Study	Research scope or contingent	Methodology	Main findings	Suggested mechanism of LAN influence	Limitations of the study
Kliukiene et al. (2003) [a follow up study to Tynes et al. (1996)]	Same as in Tynes et al. (1996) but for the period of 1961–2002, with 99 reported BC cases.	Same as in Tynes et al. (1996)	SIR for BC was 1.30 (95 % CI = 1.05–1.58), compared with the total Norwegian female population. Higher risk estimates for older population group (>50 years).	Same as in Tynes et al. (1996)	The same
Rea et al. (2011)	72 female school teachers with minimal exposure to shift work residing in the rural areas of Albany, NY and in Vermont.	Photometric measurements were made at the bedroom windows for seven consecutive days and nights and in the bedrooms of a sample of female school teachers (39 teachers participated during the spring semester (January–June 2010) and 33 teachers participated in the fall semester (September–December 2010).	The female teachers who participated in this study did not have disrupted light–dark cycles similar to those associated with rotating shift-working.	No mechanism suggested	Small sample size residing in dimly lit rural areas and limited number of measurements at different seasons of the year, resulting in limited statistical power; lack of matching controls among BC affected individuals.

(continued)

Table 11.1 (continued)

Study	Research scope or contingent	Methodology	Main findings	Suggested mechanism of LAN influence	Limitations of the study
C. Population level studies					
Cohen et al. (1983)	3,183 female patients diagnosed with BC in Israel over a 7-year period from 1960 to 1966.	Time-series analysis of monthly BC detection rates.	Peaks in BC detection tended to occur during spring through autumn.	The pattern is suggested to be of hormonal nature.	Simple analysis of monthly time series, with no linkages to LAN analyzed.
Davis et al. (2001)	813 cancer patients aged 20–74 years and 793 control subjects of comparable age.	Personal interviews conducted on sleep habits and the bedroom-light environment 10 years before diagnosis and lifetime occupational history. Conditional logistic regression used to adjust for other potential risk factors, such as family history of cancer, parity, oral-contraceptive use, and hormone-replacement therapy.	The analysis revealed that BC risk was significantly higher among subjects with sleep disturbance (OR = 1.14, 95 % CI = 1.01–1.28), but no clear association between bedroom-light intensity and BC was found (OR = 1.4, 95 % CI = 0.8–2.6).	Suppressing the normal nocturnal production of MLT potentially resulting in the release of estrogen by the ovaries.	Self-reported LAN exposure levels instead of actual measurements; potential presence of recall errors due to a possibility that reported prior sleep habits could be affected by more recent disease experience, resulting in differential recall for case patients relative to control subjects.

(continued)

Table 11.1 (continued)

Study	Research scope or contingent	Methodology	Main findings	Suggested mechanism of LAN influence	Limitations of the study
O'Leary et al. (2006)	576 female residents of Nassau and Suffolk counties of Long Island, NY, who lived in the same home for 15 years or more and diagnosed with BC in 1996–1997, and 585 population-based controls.	Personal interviews about LAN-exposure histories at work and at home. The questionnaire included questions about shift working, income, ethnicity, family health history and the number of sleep hours, frequency of turning lights-on during the night, and the length of time the light was on during the nighttime.	BC was not found to be associated with shift work (OR = 1.04, 95 % CI: 0.79, 1.38) or evening shift work (OR = 1.08, 95 % CI: 0.81, 1.44). Moreover, overnight shift workers were at lower risk than women never working shifts (OR = 0.55, 95 % CI: 0.32, 0.94). However, women who frequently turned on lights at home during sleep hours (>or = twice/week and >or = twice/night) had increased risks (OR = 1.65, 95 % CI: 1.02, 2.69).	LAN-associated circadian disruption and nocturnal MLT suppression.	The survey based on a relatively small population sample with different employment histories and did not include questions about bedroom light intensity; a possibility of potential recall error due to reliance on self-reported histories; no specific occupational health risks reported.
Kloog et al. (2008)	Female population of Israel	Linking LAN levels obtained from nighttime satellite images with breast and lung cancer incidence rates in 147 residential communities in Israel, while controlling for incomes, birth rates, population density, etc.	The study revealed a positive and statistically significant association of BC rates with LAN exposure ($P < 0.05$) and no such association for lung cancer. The study estimated 73 % higher BC prevalence in the highest lAN exposed communities compared to the lowest LAN exposed communities.	Suppression of MLT and/or disruption of clock gene function.	Difficulties to establish causality typical for population based studies, potential misclassification of exposure (e.g., misrepresentation of individual LAN exposure levels) and a lack of controlling for individual level confounders such as smoking, alcohol consumption, family history, birth history etc.

(continued)

Table 11.1 (continued)

Study	Research scope or contingent	Methodology	Main findings	Suggested mechanism of LAN influence	Limitations of the study
Kloog et al. (2009a)	Male population of 164 countries worldwide	Linking population-weighted country-level LAN levels obtained from nighttime satellite images with incidence rates of three most common cancers (prostate, lung, and colon) in men, while controlling for *per capita* income, percent urban population, and *per capita* electricity consumption.	Holding other variables at average levels across the 164 countries, the analysis yielded a risk of PC in the highest LAN-exposed countries more than doubled that observed in the lowest LAN exposed countries.	The same	The same
Kloog et al. (2009b)	Individual BC patients residing in the city of Haifa, Israel	Estimating individual-level exposure to outdoor LAN by using information on LAN emanating from municipality-maintained light poles and linking this information to residential locations of BC patients by using kernel density function tools in GIS.	BC incidence was found to be significantly associated with outdoor LAN near residences ($P < 0.01$), controlled for population densities, SO_2 air pollution and indicator of local welfare levels.	The same	The same

(continued)

Table 11.1 (continued)

Study	Research scope or contingent	Methodology	Main findings	Suggested mechanism of LAN influence	Limitations of the study
Kloog et al. (2009c)	1,679 women (794 diagnosed with BC and 885 population based controls) obtained from the "BC in Northern Israel" study, initiated in 2000.	Individual level case–control study performed using questionnaires and self-reported LAN exposure levels.	Bedroom LAN was found to be significantly associated with BC risk (OR = 1.220, 1.118–1.311), controlling for education, ethnicity, fertility, and alcohol consumption. Yet stronger LAN effect was found among Jewish women (OR = 1.278, 95 % CI = 1.115–1.414; $p < 0.001$) than among the rest of the study cohort.	Suppression of MLT and/or disruption of clock gene function.	Self-reported LAN exposure levels instead of actual measurements; potential presence of recall errors and missing confounder bias (e.g., related to the use hormone therapy, family history and occupational exposure).
Kloog et al. (2010)	Female population of 164 countries worldwide.	Linking population-weighted country-level LAN levels obtained from nighttime satellite images with incidence rates of five most common cancers in women (breast, lung, colorectal, larynx, and liver), while controlling for fertility rates, *per capita* income, percent urban population, electricity consumption and several other potential confounders.	The study found a significant positive association between population LAN level and incidence rates of BC ($P < 0.05$). However, there was no such an association between LAN and other cancer types under study. With other variables being held at their average values, the analysis yielded a 30–50 % higher risk of BC in the highest LAN exposed countries compared to the lowest LAN exposed countries.	The same	The same

(continued)

Table 11.1 (continued)

Study	Research scope or contingent	Methodology	Main findings	Suggested mechanism of LAN influence	Limitations of the study
Borisenkov and Anisimov (2011)	General data on cancer morbidity in 35 countries worldwide (1985–2007).	Cross-sectional multivariate analysis of several factors potentially affecting cancer rates.	Morbidity from breast, endometrial, colon and lung cancer shown to increase north of the equator while that from cervical, gastric and hepatic cancers found to be relatively higher in circum-equatorial populations. In the 2000s, climate-related risks of hormone-dependent cancers have gradually decreased, while those of economic and social factors have increased.	The same	The same
D. Animal models and laboratory research					
Hrushesky et al. (1979)	Human subjects	Laboratory research	A circadian rhythm was found in the level of estrogen receptors in BC in women, with higher values in late autumn and lower values in the spring.		
Lewy et al. (1980)	six human subjects	Exposing to different light intensities and testing for blood MLT levels.	Bright artificial light suppressed nocturnal secretion of MLT in human subjects. In particular, MLT concentrations decreased 10–20 min after the subjects were exposed to 2500-lux incandescent light and reached near-daytime levels within 1 h. However, room light of less intensity, which is sufficient to suppress MLT secretion in other mammals, did not lead to the same effect in humans.	Very bright light has a suppressant effect on MLT production by pineal gland.	Small sample size with potentially limited applicability to general population considering different light condition and potentially accumulating effects of light exposure during prolonged periods of time.

(continued)

Table 11.1 (continued)

Study	Research scope or contingent	Methodology	Main findings	Suggested mechanism of LAN influence	Limitations of the study
Tamarkin et al. (1982)	20 human subjects	Plasma MLT concentrations were determined over a 24 h period.	Night-time plasma MLT levels were found to be lower in women with estrogen receptor positive (ER+) BC than in ER negative BC and in healthy control women.	Low nocturnal MLT concentrations are associated with the presence of estrogen receptor positive BC.	The same
McIntyre et al. (1989)	13 human subjects	Exposure to five different light intensities with subsequent testing for blood MLT levels.	Maximum suppression of MLT following 1 h of light at midnight was found to be 71, 67, 44, 38, and 16 % with intensities of 3,000, 1,000, 500, 350, and 200 lux, respectively.	Direct MLT suppression by light.	
Anisimov et al. (1997)	Laboratory animals (two-month old outbred female LIO rats)	Case–control study with one group of rats given MLT 5 days a week during the night-time.	Incidence of carcinomas in the ascending colon in rats received MLT was significantly reduced ($P < 0.01$).	Inhibitory effect of MLT on intestinal carcinogenesis.	Animal models with unsure applicability to humans.

(continued)

Table 11.1 (continued)

Study	Research scope or contingent	Methodology	Main findings	Suggested mechanism of LAN influence	Limitations of the study
Van den Heiligen berg et al. (1999)	Laboratory animals (male Wistar rats)	The rats were randomized into 3 groups. Rats from group 1 received DEN only, while rats from group 2 also received Phenobarbital for 4 weeks as a promoting agent and rats from group 3 were exposed to continuous light. Three months after starting DEN treatment, urinary 6-sulfatoxy MLT(αMT6s) excretion, a marker of circadian clock function, was tested and laparotomy was performed.	The proportion of rats with macroscopic nodules on liver surface was 72 % (LD group), 89 % (LD pheno group) and 95 % (LL group) ($p = 0.1$). Nodules were more numerous and larger both in the LL group and in the LD-pheno group as compared to the LD group.	Light-induced circadian clock suppression exerted a cancer promoting effect.	The same
Brainard et al. (2001)	72 human subjects with the mean age of 24.5 years	Exposure to monochromatic light 2:00 and 3:30 A.M. with subsequent testing of blood samples for MLT levels.	446–477 ηm was found to be the most potent wavelength region providing circadian input for regulating MLT secretion.	MLT suppression by short light wavelengths.	

(continued)

Table 11.1 (continued)

Study	Research scope or contingent	Methodology	Main findings	Suggested mechanism of LAN influence	Limitations of the study
Wright et al. (2004)	42 human subjects	Laboratory analysis of half-hourly saliva samples after exposure to light sources of different wavelengths.	The shorter wavelengths of 470, 495 and 525 ηm showed the greatest MLT onset advances ranging from approximately 40–65 min while the longer wavelengths produced no significant phase advance.	The same	
Blask et al. (2005)	Human volunteers and laboratory animals (rats).	Venous blood samples were collected from healthy, premenopausal female volunteers during the daytime, nighttime, or nighttime following 90 min of ocular bright, white fluorescent light and perfused in rats.	Tumors perfused with MLT-deficient blood collected following ocular exposure to LAN exhibited the daytime pattern of high tumor proliferative activity. Tumor growth response to exposure to light during darkness appears to be intensity dependent and circadian MLT signal inhibits human BC growth but this effect is extinguished by short-term ocular exposure to bright white LAN.	Suppression of pineal MLT production by exposure to light at night.	

in Northern Israel, who published in 1970 the paper, entitled: *Comparison of Seasonal Influence on Pregnancy Duration in Man and Domestic Animals.*

The paper was published in the December (1970) issue of the International Journal of Biometeorology and has been largely unnoticed by the scientific community. It got only five cites in scientific papers over 40+ years, with four of them in Dr. Cohen's own publications.

Using empirical examples derived from different empirical studies (such as the observed durations of pregnancy of different animal species), and comparing these data to humans, Dr. Cohen noticed a strong correlation between the duration of pregnancy and the month of birth.

According to his observation, the observed duration of pregnancies was shortest in summer and fall and longest—in the winter months, between December and January. He attributed this association to the *duration of daylight* during different seasons of the year, to which both humans and animals are exposed. Based on this observation, Dr. Cohen suggested that seasonal patterns may, in fact, be linked not only to the duration of pregnancy, but also to the reproductive function in general, including menarche, reproductive activity, and the total number of births and stillbirths.

Eight years later, in 1978, Dr. Michael Cohen (apparently Phillip Cohen's son) published together with his colleagues, Marc Lippman and Bruce Chabner of the National Cancer Institute in Maryland another short paper in the "Hypotheses" section of the *Lancet.* As the authors acknowledged, their paper was inspired by Dr. Phillip Cohen's 1970 study and hypothesized, apparently for the first time, that increased BC risk in modern societies may be attributed to ambient lighting and potentially inhibiting effect of artificial light on MLT secretion. In particular, as Cohen et al. (1978) suggested, impaired pineal MLT secretion (attributed e.g., exposure to environmental lighting) may result in unopposed estrogen secretion and an increased susceptibility to BC.

As supporting evidence for their hypothesis, Cohen et al. (1978) cited several observations derived from previous studies, viz.: (1) Pineal calcification is commonest in countries with high rates of BC and lowest in areas with a low incidence; (2) Chlorpromazine raises serum-MLT with several studies reporting that psychiatric patients taking chlorpromazine have a lower incidence of BC; (3) Laboratory studies reported that the pineal and MLT may influence tumor induction and growth in experimental animals; (4) There are several studies suggesting a direct influence of the MLT hormone on the ovarian function, and possibly on estrogen production, and that (5) Impaired pineal secretion is believed to be an important factor triggering puberty (with early menarche being a risk factor for BC).

Nearly 15 years later, Stevens et al. (1992) published the first comprehensive review of MLT-BC studies. However, the authors of this critical survey considered two potential causes of increased BC incidence in industrialized societies: light-at-night (LAN) and electromagnetic fields (EMF)—two products of electric power, not only LAN, both of which, in their view, can potentially suppress pineal function and increase BC risk.

However, the former possibility (that MLT secretion can be influenced by EMF) was later on "dropped" as a viable cause of BC in another critical survey published 9 years later by Stevens and Rea (2001). According to the results of that survey, based on empirical evidence accumulated since the early 1970s till late 1990s, only the interaction between LAN and MLT secretion can be uphold as the only viable link, leading to the increased risk of BC ogenesis. The authors of this survey (*ibid.*) also emphasized that electric lighting is different in its spectral properties from natural light and may contribute substantially to "circadian disruption," thus becoming a major cause of "endocrine disruption" and thus increase BC incidence in industrialized countries. Another important contribution of Stevens and Rea's (2001) survey to the existing then body of literature was their hypothesis that electrical light may have adverse effects not only on MLT release but also on other physiological rhythms, including the sleep/wake cycle. This suggestion "widened" the "MLT-BC," being prevalent in earlier studies.

In a separate study published 3 years later, Nelson (2004) also raised a possibility that stress attributed to LAN can impair immune function and thus increase disease susceptibility. He also suggested that photoperiodic information, mediated by MLT, may influence immune responses.

In another survey, Reiter et al. (2007) came to a similar conclusion, citing increasing empirical evidence, from both epidemiological studies and laboratory research, that LAN has the direct capability of elevating cancer risk. As the authors of this survey acknowledged, the actual mechanism of this association is not totally clear: it may result from disruption of circadian rhythms and/or from a suppression of MLT due to LAN exposure, with both of these elements potentially contributing to cancer onset and/or progression.

Studies of Specific Population Cohorts

As we already noted, the MLT-BC&PC hypothesis puts a particular emphasis on the direct effect of LAN on human eye which transmits LAN signals to the pineal gland throng and thus inhibits MLT secretion. As a result, any person who is exposed to high LAN intensities is expected to exhibit elevated BC&PC risks, while persons less exposed to LAN are expected to be at lower BC&PC risk. This simple logic has generally been upheld by several case–control studies, published since the early 1980s to date. In particular, such studies compared cancer risks in LAN exposed individuals with those in less exposed ones. In the first population group (that is, high LAN-exposed individuals) were nighttime shift-workers, such as nurses, flight attendants and telegraph operators, while unexposed controls or the totally or partially blind were assigned to the second (that is, expectedly low BC&PC risk) group.

Thus, in an early study carried out in this domain, Hahn (1991) analyzed hospital discharge records of 11,769 women obtained from the 1979–1987 U.S. National Hospital Discharge Survey, excluding women with diabetes. The researcher compared the odds of profound binocular blindness among BC diagnosed women with the odds of profound binocular blindness among women diagnosed with coronary heart disease or stroke. As the study revealed, profoundly blind women were half as likely to have BC as women who were not profoundly blind. This effect diminished substantially with increasing age.

Several years later, Verkasalo et al. (1999) analyzed medical records of some 10,935 women with visual impairment identified from the Finnish Register of Visual Impairment and followed up for cancer through the Finnish Cancer Registry for years 1983–1996. Standardized incidence ratios (SIR) of BC for subgroup of women with different degrees of visual impairment were then calculated. As the study revealed, BC risk decreased by degree of visual impairment ($P < 0.04$), thus suggesting a dose–response relationship between visible light and BC risk.

The studies of high LAN-exposure groups are more numerous but apparently started only in the early 1990s. In one of the first studies of this type, Suvanto et al. (1993) examined 40 Finnish female flight attendants working on transcontinental flights. Measurements of oral temperature, alertness, and visual search were performed at 2 h intervals 2 days before the flight from Helsinki to Los Angeles, during the second and the fourth day in the USA and during the second and fourth day after the return flight to Finland. The study revealed the de-synchronization of the body temperature and the phases of the alertness and visual search rhythms shifting. A delay of the acrophases of the circadian rhythms was observed during the second and fourth day after the return flight. The acrophase of alertness was still delayed after the fourth day for 35 min and the acrophases of body temperature and visual search were 2 h 2 min and 3 h 8 min delayed, respectively. The dysfunctions were attributed to the disruption of daily rhythms.

Several years later, Pukkala et al. (1995) analyzed a larger cohort of Finnish flight attendants, which included 1,577 female and 187 male cabin attendants who worked for various Finnish airline companies. The authors of this survey compared standardized incidence ratio of several tumors with the expected number of cases based on national cancer incidences. Significant excess of BC (OR = 1.87; 95 % CI 1.15, 2.23) and bone cancer (OR = 15.10; 95 % CI = 1.82, 54.40) was found among female workers.[1] The risk of BC was most prominent in 15 years after recruitment. The authors attributed such access in morbidity to the adverse effects of ionizing radiation during flights.

In 2001, two other studies dealing with BC risks in high LAN exposure groups—by Hansen (2001) and Schernhammer et al. (2001)—were published. In the former study, Hansen (*ibid.*) analyzed 7,035 Danish women diagnosed with BC, of whom 497 worked at night, comparing them with individually matched controls. Employment records obtained form from pension funds, and BC odds ratios were calculated using conditional logistic regressions, adjusted for

---

[1] OR is an Odds Ratio, which is the ratio of the odds of a given health effect to occur in the exposed (case) group to the odds of the said effect to occur in the unexposed (or control) group. The OR of 1.87 thus implies that flight attendants covered by the study were 87 % more likely to develop BC than people in the general population used as controls.

socio-economic status, age at the birth of first and last child and the number of children. As the analysis indicated, the odds ratio for BC among women who worked at night at least half of a year was 1.5 (95 % confidence interval, 1.2–1.7). The study also revealed a tendency to increasing odds ratio by increasing duration of nighttime employment. The author of this study attributed these effects to adverse psychological and physiological effects associated with irregular working hours and specifically working at night.

In a separate study of 78,562 women participating in the U.S. Nurses' Health Study with 2,441 documented incidences of BC, Schernhammer et al. (2001) found a moderate increase in BC risk among women who worked on night shifts for 1–14 years (RR = 1.08; 95 % CI = 0.99, 1.18) and 15–29 years (RR = 1.08; 95 % CI = 0.90, 1.30), respectively), and higher risk among women who worked 30 or more years on the night shift (RR = 1.36; 95 % CI = 1.04–1.78). The revealed effect was attributed to the suppression of production of MLT with its potential oncostatic action due to LAN exposure.

Population Level Studies

Apparently the first known attempt to test the LAN-BC association against general population data (as opposed to data for specific population groups, singled out by its professional occupation or health status, e.g., nighttime shift workers or the visually impaired), belongs to the same person who set forward the "light cycle" hypothesis—Dr. Phillip Cohen. In 1983, he and his colleagues published an empirical paper that analyzed the time of diagnosis of 3183 female patients diagnosed between 1960 and 1966 with BC in Israel (Cohen et al. 1983). A general conclusion stemming for the results of this study was that the peaks in BC detection tend to occur during spring and autumn, leading the authors to suggest that this pattern may be of hormonal nature. No link of BC incidence to LAN was hypothesized and investigated in this study though.

Eighteen years later, Davis et al. (2001) carried out a case control study of 813 cancer patients aged 20–74 years and 793 control subjects of comparable age. Personal interviews were conducted on sleep habits and the bedroom-light environment 10 years before diagnosis and lifetime occupational history. Conditional logistic regressions were used to adjust for other potential risk factors, such as family history of cancer, parity, oral-contraceptive use, and hormone-replacement therapy. The analysis revealed that BC risk was significantly higher among subjects with sleep disturbance (OR = 1.1, 95 % CI = 1.0–1.3), but no clear association between bedroom-light intensity and BC was found (OR = 1.4; 95 % CI = 0.8–2.6).

In another population-based study published by O'Leary et al. (2006), known as the "Long Island study", the researchers investigated 576 women—residents of Nassau and Suffolk counties on Long Island, NY, who lived in the same home for more than 15 years and diagnosed with BC in 1996–1997. The BC cases were matched with 585 population-based controls. Personal interviews about LAN-exposure histories at work and home were also conducted. The questionnaire included questions about shift working, income, ethnicity, family health history and the number of sleep hours, frequency of turning lights-on during the night, and the length of time the light was on during the nighttime. According to the results of this study, BC risk was not found to be associated with either bedroom intensity bedroom light intensity (OR = 1.4; 95 % CI = 0.8–2.6) or shift work ($P > 0.1$). However, women who frequently turned on lights at home during sleep hours exhibited increased BC risks (OR = 1.65, 95 % CI: 1.02–2.69).

Several years later, Kloog et al. (2009a), investigated the association between bedroom light levels and BC in Israel, using a similar questionnaire such as that which was used by O'Leary et al. (2006). Kloog et al.'s study covered 1,679 women, including 794 BC diagnosed women and 885 population-based controls. The study was performed using questionnaires and self-reported LAN exposure levels—ranging from 0 (absolutely dark) to 4 (highly illuminated). Bedroom LAN was found to be significantly associated with BC risk (OR = 1.2, 1.1–1.3), controlling for education, ethnicity, fertility, and alcohol consumption. Yet, stronger LAN effects on cancer risk were found among Jewish women (OR = 1.3, 95 % CI = 1.1–1.4), than among the rest of the study cohort.

Kloog et al. (2009a) study was thus the first large scale case–control study of the general population that identified a significant positive association between bedroom light levels with BC risk, providing evidence that the relative risk of BC appears to increase significantly in more illuminated sleeping environments. The main conclusion the authors drew from their study was that only the working environment, as noted in shift workers studies, should be considered a potential BC risk factor, but also the modern human "sleeping habitat" with often high LAN levels should be considered as a BC risk factor as well.

Kloog et al. (*ibid.*) also discussed differences in their results with those obtained in previous population level studies which did not identify the link between BC risk and LAN in the sleeping habitat (cf., e.g., Davis et al. (2001) and O'Leary et al. (2006)). In particular, two explanations were proposed. First, the former's study population was relatively homogenous, with a separate analysis performed for Jewish women, while in the study by Davis et al. (2001) involved a more heterogeneous general population of diverse ethnic background. Second, the authors of the former survey drew attention to the fact that the studies under comparison were conducted some 15 years apart, with the study by Davis and co-authors (*ibid.*) was conducted in 1992–1995 and the study of Kloog et al. in 2008–2009. Since the early 1990s, light pollution has increased, and women are currently exposed to higher light intensity levels. Lastly, in the past decade, light bulbs emitting bluer light waves ($\sim$460 ηm) have been widely introduced, to save energy consumption and reduce CO_2 emission. The introduction of these new artificial light sources may have had adverse health effects. This conclusion coincides with the results of a study carried out by Cajochen and colleagues (2005) and another study by Wright et al. (2004) which revealed that short wave length light (of 470–525 ηm wave length) tend to decrease MLT production while increasing alertness, body temperature, and heart rate. Concurrently, exposure to a wave length of 550 ηm for the same duration of time and intensity level did not cause such effects, as discussed elsewhere in this book.

During the short period of 2008–2010, three more large-scale population-based studies were published (Kloog et al. 2008, 2009b, 2010). These studies, headed by the authors of this book, will be discussed in some detail, in the next chapter.

In a separate study, published 1 year later, Borisenkov and Anisimov (2011) used general data on cancer morbidity in 35 countries worldwide (1985–2007) applying to them cross-sectional multivariate analysis of factors potentially affecting cancer rates. The analysis revealed that morbidity attributed to breast, endometrial, colonic and lung cancer tended to increase north of the equator, while cervical, gastric and hepatic cancer incidence rates were found to be relatively higher in circum-equatorial populations. The study also revealed that in the 2000s, climate-related risks of hormone-dependent tumors slumped down while those of economic and social factors increased, which is generally in line with Kloog et al.'s (2009a) findings.

Laboratory Experiments and Animal Models

The (apparently) earliest study dealing with the LAN-BC association was carried out by Hrushesky and colleagues and published in 1979 (Hrushesky et al. 1979). According to the results of this study, circannual rhythm was found in the level of estrogen receptors of BC-diagnosed women, with higher values in late autumn and lower values in the spring. This study was followed by another landmark study published by Lewy et al. 1980, who investigated six human subjects exposed to different light intensities and tested for blood plasma MLT levels. According to the results of this study, bright artificial light suppressed MLT nocturnal secretion in human subjects. In particular, MLT concentrations decreased 10–20 min after the subjects were exposed to 2500-lux incandescent light and reached near-daytime levels within 1 h. However, room light of less intensity, which is sufficient to suppress MLT secretion in other mammals, failed to do so in humans. This study thus showed for the first time that very bright light could suppress MLT in humans, while until the publication of this paper it was thought that humans were refractory to any light-induced MLT suppression.

Another study by McIntyre and colleagues (1989) confirmed Lewy et al.'s (1980) results and helped to gain additional insights on the phenomenon of MLT suppression by artificial light. The authors of this research exposed 13 human subjects to five different light intensities and subsequently tested them for blood plasma MLT levels. Maximum suppression of MLT following 1 h of light at midnight was 71, 67, 44, 38, and 16 % with intensities of 3,000, 1,000, 500, 350, and 200 lux (lx), respectively. The study not only confirmed the phenomenon of direct MLT suppression by light, but also indicated that bluer light length waves are capable of suppressing MLT more than longer light waves.

In a separate study by Tamarkin et al. (1982) tested plasma MLT concentrations in 20 human subjects over a 24 h period. Night-time plasma MLT levels were found to be lower in women with estrogen receptor positive (ER+) BC than in ER negative (−) BC and in healthy control women, thus demonstrating that low nocturnal MLT concentrations are associated with the presence of estrogen receptor positive BC.

Several years later, Anisimov et al. (1997) carried out another important study on laboratory rats. In this case–control study one group of rats was administrated with MLT 5 days a week during the night-time, and another group was not. Incidence of carcinomas in the ascending colon in rats received MLT was significantly reduced ($P < 0.01$). This study thus demonstrated, for the first time, the inhibitory effect of MLT on intestinal carcinogenesis.

Two years later, Van den Heiligenberg et al. (1999) reported the results of a similar experiment, in which laboratory rats were randomized into 3 groups. While rats from group 1 received DEN only, the rats from group 2 also received Phenobarbital for 4 weeks as a promoting agent, and rats from group 3 were exposed to continuous light. Three months after starting DEN treatment, urinary 6-sulfatoxy MLT (αMT6s) excretion, a marker of circadian clock function, was tested and laparotomy was performed. The proportion of rats with macroscopic nodules on liver surface was 72 % (LD group), 89 % (LD pheno group) and 95 % (LL group) ($p = 0.1$). Nodules were more numerous and larger both in the LL group and in the LD pheno one as compared to the LD group. The results of this study thus demonstrated that light-induced circadian clock suppression exerted a cancer promoting effect, thus completing another missing link in the LAN-BC relationship.

The work of Brainard et al. (2001) shed more light on human response to LAN, by demonstrating the effects of exposure to monochromatic light between 2:00 and 3:30 A.M. with subsequent testing of blood samples for MLT levels. The study included 72 human subjects with the mean age of 24.5 years. The results of this study were consistent with those of McIntyre et al. (1989) who also demonstrated that light of different wavelength had different effect on MLT suppression. In particular, as Brainard et al. (2001) found, 446–477 ηm was the most potent wavelength region providing circadian input for suppressing MLT secretion. In a later study, Wright et al. 2004 also found that the shorter wavelengths of 470, 495 and 525 ηm showed the greatest delays in MLT secretion, ranging from approximately 40–65 min, while the longer wavelengths produced no significant phase advance.

In another pioneering study by Blask et al. (2005), venous blood samples were collected from healthy, premenopausal female volunteers during either the daytime, nighttime, or nighttime following 90 min of ocular bright, white fluorescent light. The blood samples were then perfused in rats. Tumors perfused with MLT-deficient blood collected following ocular exposure to LAN exhibited the daytime pattern of high tumor proliferative activity. Tumor growth response to light exposure during darkness appears to be intensity dependent and circadian MLT signal inhibits human BC growth but this effect is extinguished by short-term ocular exposure to bright, white LAN.

In Summary

The 30 years of research thus helped to build the following causality chain: LAN (and especially its short-length frequencies) has a suppressive effect on MLT secession (Lewy et al. 1980; McIntyre et al. 1989; Brainard et al. 2001), while MLT was demonstrated to have an inhibitory effect on carcinogenesis (Anisimov et al. 1997; Heiligenberg et al. 1999; Blask et al. 2005; Haim et al. 2010). In parallel, epidemiological level studies demonstrated the existence of the LAN-BC&PC link for both population subgroups, stratified by their occupation and/or health status (Hahn 1991; Verkasalo et al. 1999; Suvanto et al. 1993; Hansen 2001; Schernhammer et al. 2001), for sub-groups of general population (Davis et al. 2001; O'Leary et al. 2006; Kloog et al. 2009a) and for general population at large (Kloog et al. 2008, 2009b, 2010).

Part III
Light Pollution and its Potential Links to Breast and Prostate Cancers (BC&PC)

Chapter 12
Geographic Patterns of Breast and Prostate Cancers (BC&PC) Worldwide

Abstract Geographic patterns of BC&PC worldwide are surprisingly similar, with higher rates of both cancers observed in developed countries and lower rates elsewhere. The incidence rates of these cancers are also higher in more extreme latitudes, suggesting a possibility that they are related, among other factors, to LAN, considering that in extreme latitudes artificial light is often used to compensate for a shortage of natural illumination. Our analysis supports such a possibility, especially for PC.

Keywords Geographic pattern · BC&PC · Incidence rates · LAN · Vitamin D hypothesis · Electricity consumption

Geographic patterns of different diseases tend to be similar if these diseases share similar environmental risk factors and etiology.

Research hypothesis (Portnov et al. 2008)

Introduction

Although breast cancer (BC) and prostate cancer (PC) incidence rates vary considerably by country, their geographic patterns are essentially similar, as Figs. 12.1 and 12.2 demonstrate. In particular, the highest rates of both cancers are found in developed countries—North America, Western Europe and Australia,—and lower rates elsewhere, especially in Africa and most of Asia.

The association between incidence rates of these cancers and the countries' development levels, as measured by their *per capita* incomes, is indeed positive and highly significant, being especially strong for BC (see Fig. 12.3). In particular, as Fig. 12.3 shows, the rates of both cancers increase with local incomes, indicating that as a country develops its population tends to exhibit higher rates of these malignancies.

A. Haim and B. A. Portnov, *Light Pollution as a New Risk Factor for Human Breast and Prostate Cancers*, DOI: 10.1007/978-94-007-6220-6_12,

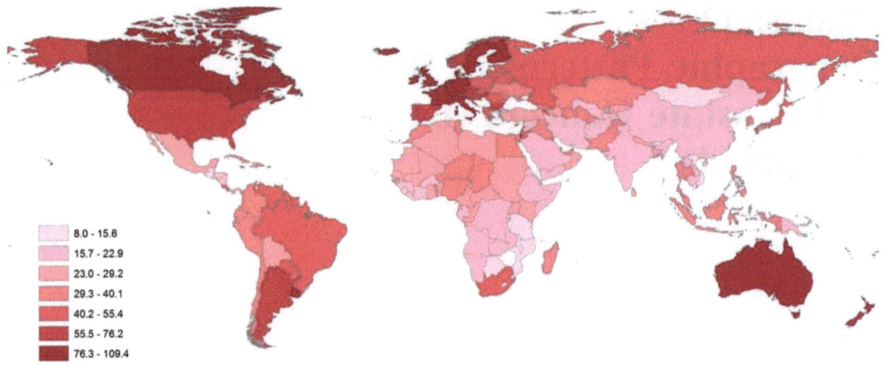

Fig. 12.1 Breast cancer (BC) rates worldwide (per 100,000). *Note* mapped using data from Globocan (2008)

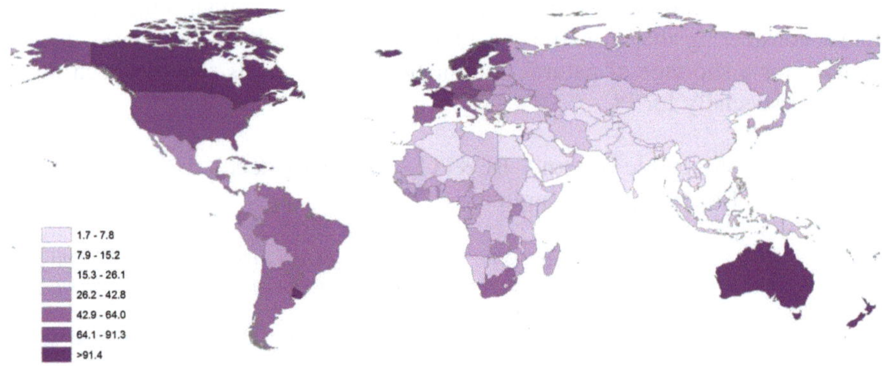

Fig. 12.2 Prostate cancer (PC) rates worldwide (per 100,000). *Note* mapped using data from Globocan (2008)

Notably, *per capita* gross domestic product (GDP) alone explains the lion share of worldwide variations of country-specific rates of BC&PC—that is, about 66 % of variance in BC rates ($R^2 = 0.656$), and nearly 40 % of worldwide variance in PC rates ($R^2 = 0.398$; see Fig. 12.3). Such a strong association is almost proverbial: *tell me how wealthy the country you live in is, and I will tell how big your chances to get sick are.*

Yet, we should acknowledge that *per capita* GDP is a very general measure of population welfare. As any good indicator, it reflects many more aspects of development than it actually measures. For instance, high *per capita* GDP is likely to be associated with higher quality of medical services and thus with a greater possibility of early detection of various malignancies.

The very quality and completeness of national cancer registries may also depend on a country's welfare levels and resources it may devote to maintaining cancer registries. As a result, *per capita* GDP is likely to act as a proxy for such registries' completeness and reliability. In particular, in a developing country, with

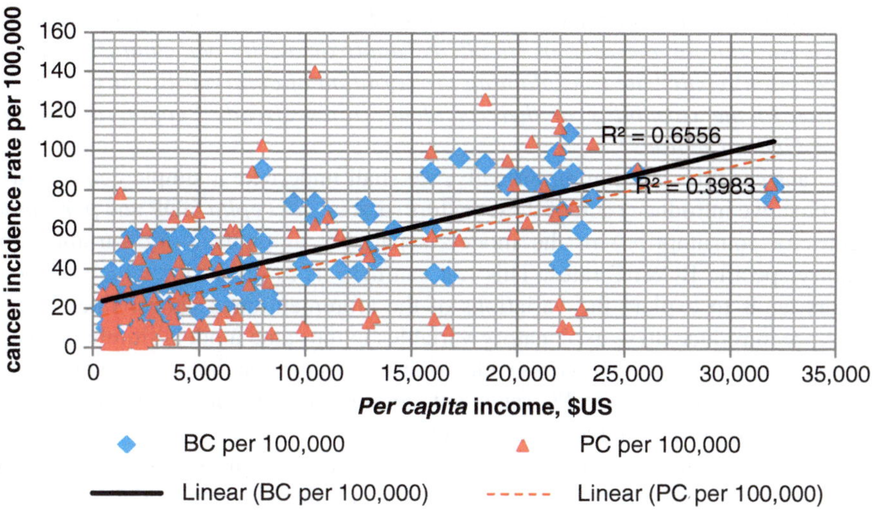

Fig. 12.3 The association between prostate cancer (PC) and breast cancer (BC) incidence rates and *per capita* GDP ($US) in 164 countries worldwide. *Note* Diagrammed using data from Globocan 2008 and World Bank (2010)

meager resources it is likely to possess, maintaining a good cancer registry may be not the top priority, compared to more urgent needs, such as security, housing, clean water and food, and provision of basic health services and vaccination. In addition, even if we assume that reported cancer rates are reasonably accurate, *per capita* GDP is likely to be strongly and positively associated with well-known risk factors of BC&PC, including local reproductive patterns, the use of oral contraceptives, hormone therapy, breast-feeding, alcohol consumption, prevalence of obesity, and a lack of physical activity. In other words, *per capita* GDP may act as a proxy for some or most of them.

Lastly, the increased use of artificial illumination in the past decades, with its suppression effect on pineal MLT production and daily rhythm disruption, which, as we hypothesize, are largely responsible for rising rates of hormone-dependent cancers worldwide, may also be positively associated with local development levels, measured by national GDPs. Indeed, in high resource countries, electricity networks are more developed than in the developing world (see e.g., Chap. 7), and electricity costs are more affordable for local residents, relative to their disposable incomes. The lower relative costs of electricity, the less restricted its use is likely to be, especially for lighting, which is less costly, compared to e.g., space heating and cooling.

In developed countries, artificial illumination is also used more widely for lighting public spaces, traffic nodes (including intercity and inner-city roads), commercial areas and sport facilities, which networks are also more dense and developed than in the less developed world. All this may thus contribute to a strong association between *per capita* GDP and artificial light.

The question thus becomes *whether will the spread of artificial illumination (which we can, for the sake of simplicity, approximate by per capita electricity consumption) contribute to explaining the variation of observed BC&PC rates, in addition to the variability of their incidence rates already "captured" by per capita GDP of nations*?

To answer this question empirically, we shall undertake a simple econometric test. First, we introduce each of these factors (that is, *per capita* GDP and *per capita* electricity consumption) in the models separately as explanatory variables for country-specific BC&PC rates. This test can help us to determine which of these factors tends to explain the variations of the aforementioned cancer rates better. Next, we introduce these explanatory variables in the same models simultaneously, to determine whether *per capita* electricity consumption helps to explain any additional variation of BC&PC rates, not captured by national *per capita* GDPs.

The results of the analysis are reported in Table 12.1.

Characteristically, *per capita* GDP and electricity consumption explain the variation of BC&PC nearly equally well when introduced in the models separately (Models B-1 and B-2, and Models P-1 and P-2; $P < 0.01$). In general, this is not surprising, considering that both factors may function as proxies for development levels of nations overall, thus being associated with other cancer risk factors, associated with economic development, such as obesity, lack of physical activity, *etc.*

However, when introduced together in the same models (Models B-3 and P-3; see Table 12.2), *per capita* GDP emerges as the strongest determinant of both cancers, while electricity consumption retains its significance in the BC model only ($P < 0.01$). This indicates that, national *per capita* electricity consumption helps to explain an additional variation of country-specific cancer rates, not captured by *per capita* GDP.

Table 12.1 Relative contributions of *per capita* GDP and electricity consumption to explaining the variation of country-specific BC&PC rates (Dependent variables—cancer rates per 100,000; method—OLS regressions)

Variable	Breast cancer (2008)			Prostate cancer (2008)		
	B-1	B-2	B-3	P-1	P-2	P-3
(Constant)	22.497	41.904	26.768	15.145	34.406	18.382
	(15.581**)	(31.451**)	(14.519**)	(6.202**)	(17.009**)	(5.720**)
GDP pc	0.003	–	0.002	0.003	–	0.002
	(17.562**)	–	(9.988**)	(10.355**)	–	(6.066**)
Electricity pc	–	8.358	2.751	–	8.021	2.084
	–	(12.293**)	(3.544**)	–	(7.770**)	(1.540)
No of cases	164	164	164	164	164	164
R^2	0.656	0.483	0.681	0.398	0.271	0.407
R^2-adjusted	0.654	0.479	0.677	0.395	0.267	0.400

Note *t*-values are in the parentheses; ** indicates a two-tailed 0.01 significance level

Table 12.2 Latitude and *per capita* electricity consumption as explanatory variables for BC&PC rates worldwide (dependent variables—country-specific cancer rates per 100,000; method—OLS regressions)

Variable	BC (2008)			PC (2008)		
	B-2	B-4	B-5	P-2	P-4	P-5
(Constant)	41.904	16.972	30.791	34.406	14.256	31.108
	(31.451**)	(6.482**)	(10.014**)	(17.009**)	(3.598**)	(6.370**)
Latitude (abs)	–	0.845	0.392	–	0.669	0.116
	–	(10.232**)	(3.972**)	–	(5.352**)	(0.742)
Electricity pc (ln)	8.358	–	6.005	8.021	–	7.322
	(12.293**)	–	(6.822**)	(7.770**)	–	(5.238**)
No of cases	164	164	164	164	164	164
R^2	0.483	0.393	0.529	0.271	0.150	0.274
R^2-adjusted	0.479	0.389	0.523	0.267	0.145	0.265

Note t-values are in the parentheses; ** indicates a two-tailed 0.01 significance level

Sunlight: Vitamin D Hypothesis

In several epidemiological studies, the deficit of sunlight was implicated in increasing BC&PC incidence rates, especially in northern countries, which is consistent with the "vitamin D" hypothesis (see *inter alia* Gorham et al. 1990; Colli et al. 2006). According to this hypothesis, vitamin D may play a pivotal role in reducing the risk of both BC&PC, while the lack of exposure to ultraviolet sunlight can increase the prevalence of vitamin D deficiency, and may thus place predisposed populations at higher risk of developing BC&PC.

Indeed, when we take a close look at the geographic patterns of BC&PC worldwide (see Figs. 12.1, 12.2), it appears that countries in both hemispheres, located closer to the poles and farther from the equator, tend to exhibit higher rates of both cancers, as compared with countries located near the equator and in more temperate latitudes. Indeed, many countries which exhibit the highest incidence rates of these cancers are located in northern and southern latitudes (e.g., Ireland, Canada, Scandinavian countries, Australia, New Zealand and the UK), whereas countries with relatively low BC&PC rates are mostly positioned between the Tropic of Capricorn and the Tropic of Cancer, on the both sides of the equator (see Fig. 12.1, 12.2).

Such a pattern indeed appears to be fully consistent with the "vitamin D" hypothesis. However, this geographic pattern does not reject the "artificial light" hypothesis either, according to which the widespread use of artificial illumination (with its adverse effects on human daily rhythms and immune system) may become a significant risk factor for BC&PC. The matter is that that the amount of solar radiation and the amount of artificial light (LAN) are likely to be inversely related. In particular, in extreme latitudes artificial light is likely to be used more often than in more temperate climates, helping to compensate for a shortage of natural illumination during long nights and seasons of the year when days are short, sky is overcast and natural light is unavailable. Even though people in such extreme latitudes may not necessarily be exposed to a short wavelength artificial illumination which suppresses MLT directly, they may produce less MLT at night because serotonin, from which MLT is produced, is not accumulated during the light phase.

So which of these two factors—natural light or LAN,—may provide a better explanation for the geographic variation in the country-specific rates of the two cancers under study?

To answer this question (or at least to get some indications about how it can be answered) we run another empirical test. In particular, we analyze the association between the rates of these cancers and two simple measures, linked to the availability of natural light (vitamin D hypothesis) and artificial light (LAN hypothesis; see Chap. 5 and 11), respectively. These simple measures are, *country's latitude* and *per capita electricity consumption*, respectively.

Both measures are, of course, relatively crude. Although latitude is directly related to the amount of sunlight available, some "representative" latitude taken for a country as whole and estimated for the center of its land mass (i.e., its geographic *centroid*) may be only a rough approximation of the country's geographic location, especially for large countries, such as Russia, USA., Canada or China. Yet, it may be more accurate for smaller countries though.

By the same token, *per capita* electricity consumption is *not* a perfect measure of LAN either. The matter is that the amount of electricity used for illumination of homes and workplaces is not identical, by any means, to the total electricity consumption in the nation, considering that large shares of electricity consumption are used for industrial production, transportation, cooling and heating of indoor spaces.

Nevertheless, if we assume that the amount of electricity used for artificial lighting is fairly proportional to the total domestic electricity consumption, the latter measure, albeit relatively crude, may be used, at least as indications, for our hypothesis's testing.

As previously, we perform out empirical test in two stages: first, we introduced the latitude and electricity consumption in the regression models separately, so as to see which of them tends to explain the variations of the cancer rates under study better. Next, we introduce the above factors in the models simultaneously, so as to determine whether *per capita* electricity consumption (the LAN hypothesis) may explain the variations of BC&PC rates better than latitude (i.e., the vitamin D hypothesis).[1]

The results of the analysis are reported in Table 12.2.

As Table 12.2 shows, *both* latitude and *per capita* electricity consumption are positively and significantly associated with BC&PC rates when introduced in the

[1] It is to be noted that in the modeling we tested different functional forms of the models and opted to use the absolute values of latitudes instead of actual (positive and negative) values, so as to represent equally extreme latitudes in both hemispheres. However, for brevity's sake we skip these technical trials and will report only the models providing the best fits and generality.

models *separately* ($P < 0.01$; Models B-2,4 and P-2,4). However, when introduced *simultaneously*, the electricity consumption emerges as the strongest predictor in both BC and PC models. Characteristically, in the PC model (Model P-5; Table 12.2), latitude is statistically insignificant ($P > 0.5$), implying that it contributes little as an explanatory variable for the variation of PC rates worldwide. Electricity consumption also emerges as a stronger predictor in the BC model (Model B-5), *albeit* both predictors (that is, *per capita* electricity consumption and latitude) are statistically significant in this model ($P < 0.01$) and, expectedly, exhibit positive associations with country-specific BC rates.

Concluding Remarks

In a recent study of spatial patterns of several malignancies in Israel, Portnov et al. (2008) hypothesized that geographic patterns of different diseases tend to be similar if these diseases share similar environmental risk factors and etiology. This hypothesis appears to be validated, when worldwide geographies of BC&PC are examined. While both of these cancers appear to be strongly and positively linked to population welfare of local residents, *per capita* electricity consumption (directly related to the amount of artificial illumination locally available) helps to explain an additional variation of the observed rates of these cancers, not explained by local welfare levels.

The effect of environmental variables on BC and PC is a well-known phenomenon. Thus, for instance, the rates of BC are quite low in Japan (Bray et al. 2004), apparently due to the local diet being rich in anti-oxidants. However, in Japanese women living in the USA, BC rates are quite similar, or even higher, than among other population groups (Miller et al. 2008), thus indicating that environmental variables, food habits and/or exposure to LAN may play a major role in increasing BC risks.

Yet, the question is *what environmental attributes may affect BC&PC risk stronger—those related to natural conditions*(e.g., *latitude*) *or those attributed to development*, viz. *electricity consumption*?

As the present analysis indicates, *per capita* electricity consumption performs better than latitude in explaining the cross-country variation of BC&PC rates. This observation is fully in line with our initial research hypothesis, according to which LAN, which is often used to compensate for a shortage of natural illumination in extreme latitudes, may increase cancer risks in some populations, due to MLT suppression and daily rhythm disruption causality chains, discussed elsewhere in this book.

Chapter 13
Light Pollution and its Associations with BC&PC in Population-Level Studies

Abstract Due to the rapid spread of electric illumination, our world is not as dim at night as it was 130 or even 70 years ago. Nightlights of constantly increasing intensity and bluer in color, which illuminate our streets and public spaces today, are captured by satellite sensors and transmitted to the earth as digital images. In this chapter we discuss two case studies, which attempted to link the above digital maps of nighttime illumination with BC&PC incidence rates, thus helping to demonstrate a potential association between them.

Keywords Nighttime illumination · Light pollution · Breast and prostate cancers · Satellite images · Inter-country differences · Multivariate analysis · Geographic information systems (GIS) · Population-weighted LAN exposure · Effect estimates

> *Lighten the darkness of the night with the light of day. The day is approaching that is neither day nor night.*
>
> Passover song

Dark-Less World

500, 200, 130 and even 70 years ago, the whole world was almost totally dark at night. Only major cities, such as Paris and London, were dimly lit by whatever meager means available—oil lamps, torches, and, more recently, by kerosene lamps and gas-burning lanterns. The situation has changed dramatically with a wide and rapid spread of electricity for lighting, the process which we discuss in Chap. 7 of this book.

Nighttime satellite photometry, collected in the framework of the U.S. Air Force Defense Meteorological Satellite Program—Operational Linescan System (DMSP-OLS), has been in place for more than a decade, although the quality and resolution of satellite images have been constantly improving (Cinzano et al. 1999; Elvidge et al. 2009a, b, c; Baugh et al. 2011). The DMSP-OLS was designed to

A. Haim and B. A. Portnov, *Light Pollution as a New Risk Factor for Human Breast and Prostate Cancers*, DOI: 10.1007/978-94-007-6220-6_13,

collect global LAN imagery using a broad spectral band of both visible and thermal lights. The radiance calibrated maps of nighttime illumination with a resolution of about 2.7 km^2(1.64 × 1.64 km) per one digital pixel are available to researchers.

Over the years, DMSP-OLS nighttime satellite imagery has been used for mapping sky brightness and built surfaces, construction of global "poverty maps", estimation of ecological footprints of different countries and country-specific electrification rates, spectral identification of lights, monitoring forest fires, and for many other studies and development tasks (cf. *inter alia* Cinzano et al. 1999; Elvidge et al. 2009, 2010a, b; Doll, 2011; Baugh et al. 2011; Small et al. 2011).

However, the idea to link the digital satellite maps of nighttime illumination (such as that featured in Fig. 13.1) with place specific incidence rates of BC&PC was originated in the research headed by the authors of this book and published in several scientific papers.

In this chapter, two of these studies (Kloog et al. 2009b, 2010), their results and interpretations are discussed in turn, first, the study of LAN-PC association in men, and then, LAN-BC association in women.

Our research hypothesis was relatively simple: If there is a *significant relationship between population LAN exposure and BC&PC incidence rates, then there should be a significantly strong association between LAN intensities and BC&PC, but not with other cancers, such as colon, larynx, lung,* etc.

Fig. 13.1 LAN intensity in Europe and the north of Middle East (*Note* The map features polygons of relative light intensity constructed using 1997/98 DMSP-OLS nighttime satellite imagery)

Case Study 1: Global Co-distribution of LAN with Prostate Cancer in Men[1]

According to the World Health Organization, PC is the most common cancer in men, with some 245,000 new cases added annually worldwide (ACS 2007). In Bangladesh, its age standardized rate (ASR) does not exceed 0.3 per 100,000, while in the U.S.A., it reaches as many as 124.8 cases per 100,000. Although several population-level studies of cancer risk factors have been carried out, most of these studies were focused on specific, such as breast and liver cancers) and mainly among women (Bray et al. 2004; Althuis et al. 2005; Bonner et al. 2005; Brody et al. 2007; Kloog et al. 2008).

The data on cancer ASRs in men were obtained from the GLOBOCAN 2002 database, maintained by the IARC (Parkin et al. 2001a, b, c; 2005). ASR is a summary measure of a rate that a world population would have if it had a standard age structure (Parkin et al. 2005) and is calculated per 100,000 per year. The data were also obtained from the same source for the other common cancers in men—lung, and colon,—in order to serve as negative controls.

Explanatory Variables for Country-Specific Cancer Rates

LAN emission. As discussed elsewhere in this book, the LAN variable was used to explore whether country-specific cancer rates tend to increase with increase in LAN levels. Using simple average country LAN exposure estimates may bias the results, due to countries' differences in geography and population structure. For instance, if simple country-wide LAN averages are calculated, large and unevenly populated countries, such as Canada and Australia, are likely to exhibit very low average levels of LAN. To minimize this bias, we used a novel method of calculating LAN exposure, which took into account both a country's geographic distribution of population and its local LAN intensities (see the subsection on GIS analysis for more detail).

GDP per capita ($US) is a commonly used measure of population welfare which reflects differences in the diet and life-styles of different socio-economic strata (Hulshof et al. 1991, 2003). As several previous studies indicated, cancer rates tend to be higher among high income than across low income strata and are significantly higher in the developed than in developing countries (Bray et al. 2004; ACS 2007).

Percent of urban population. Living in cities is often associated with a considerable amount of physiological stress as a consequence of high residential

[1] The subsection is based on: Kloog I., Haim, A. Stevens R.G. and B.A. Portnov, The Global Co-Distribution of Light at Night (LAN) and Cancers of Prostate, Colon and Lung in Men, *Chronobiology International*, 2009, 26(1): 108–125.

densities, traffic congestion, and air pollution, which may increase cancer risk (Han and Naeher 2006). In addition, residents of urban areas are exposed to more environmental, second-hand smoking, due to high residential densities, which is another cause of e.g., lung cancer (Volzke et al. 2006). Dietary differences and reduced physical activities associated with urban living may also play a role in the development of cancer.

Electricity consumption (kWh per capita). Electricity consumption may be an indicator of socio-economic development and industrial emission of gaseous substances associated with electricity production (Gram-Hansenn and Petersen 2004; Jumbe 2004). *Per capita* electricity consumption may also serve as a proxy for electromagnetic field exposure (EMF) exposure, which although controversial as a causal agent, could be a possible risk factor for the development of cancer (Demers et al. 1991; Davis et al. 2006; Roosli et al. 2007).

Regional indicators. PC distribution map (see e.g., Fig. 12.2 in Chap. 12) reveals clusters of countries with similar PC rates (e.g., Western Europe, East Asia and Middle East are clear example of such clustering). In order to take this clustering into account, several dichotomous variables (termed "regional indicators") were included in the analysis as additional predictors: Middle East, Africa, East Europe, Asia, South America, and Middle Asia. Each indicator takes on value 1 if a country is located in a particular geographic region and zero otherwise (such as in North America, Australia, and Western Europe).

Data for the present analysis were obtained from the two main sources:

- Country-level data on *per capita* GDP, percent urban population, and *per capita* electricity consumption for 1998–1999 were obtained from the ESRI ArcGIS database and the CIA World Fact Book (CIA 2006; ESRI 2007);
- Data on nighttime illumination (LAN) were obtained from the U.S. Defense Meteorological Satellite Program (DMSP 2004). The DMSP satellite provides continuous reading of the entire Earth surface during the nighttime as it cycles around the globe. The satellite image for 1996/97, used in the analysis, was generated by DMSP by averaging daily readings of the satellite sensors and removing cloud cover.

GIS Analysis

In recent years, Geographic Information Systems (GIS) have become an important research tool for cancer-related studies (Krieger et al. 2002; Maheswaran et al. 2002; Scott et al. 2002; Banerjee et al. 2003; O'Leary et al. 2004; Kloog et al. 2008). In these studies, GIS is used to calculate the distance between residences and hazardous waste sites, as well as to account for the spatial clustering and variation of cancer cases and to capture the spatial and temporal heterogeneity in survival patterns. In the study reported in this chapter, GIS technology was used for matching country-specific cancer rates with the LAN levels obtained from satellite images. The task was performed using the "spatial join" tool in the

ArcGIS™ software, which combines data from two geographic layers by appending attributes from one layer to another, based on the relative location of features in the layers (Minami and ESRI 2000).

The "spatial join" was performed in two steps. In the first step, a worldwide radiance-calibrated satellite image of nighttime illumination, which reports average nightlight intensity in 1996/97 measured in light radiance units (i.e., $nW/cm^2/sr$), was imported to the ArcGIS™ software. The image reflects a fraction of light escaped into the space and detected by the satellite's sensors. Although these satellite measurements are magnitude lower than actual LAN levels detected on the ground, they represent accurately the relative levels of nightlight intensity observed in the localities (DMSP 2004). The original image size was about 43,200 by 21,600 pixels.

It should be noted that geo-statistical tools available in GIS for working with raster (pixel) images are rather limited. Therefore, for the consequent analysis the original nighttime illumination image were converted into a vector map using the ArcGIS™ "raster-to-feature" conversion tool. The conversion resulted in a *vectorized* map containing approximately 3,800,000 polygons characterized by various LAN intensities (with a minimum LAN value of 0, no illumination, and a maximum value of 255 $nW/cm^2/sr$, which is maximum illumination).

The polygon layer (map) obtained thereby was then overlapped with a map showing the location of all major populated places (>1,000 residents) of the world obtained from the Geonames database (Geonames 2008). At the next step, average LAN values were calculated for each populated place *i*, by obtaining LAN values from the LAN intensity polygon into which the populated place falls. The locality-specific LAN values (LAN_i) obtained thereby were then multiplied by the population size of localities (POP_i) and summed for each country *j* under study. Next, these summary values were divided by the total population size of the country's populated places (ΣPOP_{ij}) to obtain the average LAN exposure estimate per person ($\overline{LAN_j}$) in each country (*j*) under study:

$$\overline{LAN_j} = \sum_{i=0}^{n} LAN_{ij} \times POP_{ij} \Big/ \sum_{i=0}^{n} POP_{ij}$$

where *n* is the total number of populated places in country *j*.

Statistical Analysis

To identify and measure the significance of factors affecting the selected cancer rates, the following linear regression model was used:

Cancer incidence rate = B0 (constant) + B1* (Electricity consumption) + B2*(GDP *per capita*) + B3*(LAN) + B4*(Percent of Urban population) + B5*(Middle East) + B6*(Africa) + B7*(East Europe) + B8*(Asia) + B9*

(South America) + 10*(Middle Asia) and +ε (random error term), where B0, …, B10 are regression coefficients. [During the analysis several other functional forms of the model (e.g., log-linear and double-log forms) were tested, and only the results of the best performing (linear) model are discussed below].

The residuals of the OLS model were tested for the presence of spatial autocorrelation using the Moran's I test statistic. The test showed significant clustering of residuals (Moran's Indicator: 0.270–4.372, $P < 0.001$) which necessitated the use of spatial dependency (SD) models to take the spatial dependency of residuals into account and improve the robustness of regression estimates (Anselin 1999). The Spatial error regression analysis was performed using the GeoDa™ spatial analysis software. However, the resulting models were found to be essentially similar to OLS results and, therefore, are not reported in the following discussion for brevity's sake.

Factors Influencing PC Rates

Figure 13.2 reports statistical significance of selected variables affecting male cancers' rates (as estimated by *t*-statistics from multivariate regression models). Interpretation of this graph is fairly straightforward: Absolute values of *t*-statistic greater than 1.98 indicate a 0.05 (two-tailed) significance level, while values >2.61 indicate a 0.01 (two-tailed) significance level (that is, a probability of error in the estimate, 0.01 = 1 %).

As expected, *per capita GDP* is positively associated with ASRs in all the models ($P < 0.01$). Only *lung cancer* appears to exhibit a significantly positive association between ASR and percent country's *urban population*. Concurrently, among the three cancers analyzed, only PC exhibits a significant positive correlation with both *LAN exposure* ($t = 2.92$; $P < 0.01$) and *per capita* electricity consumption ($t = 3.68$; $P < 0.01$).

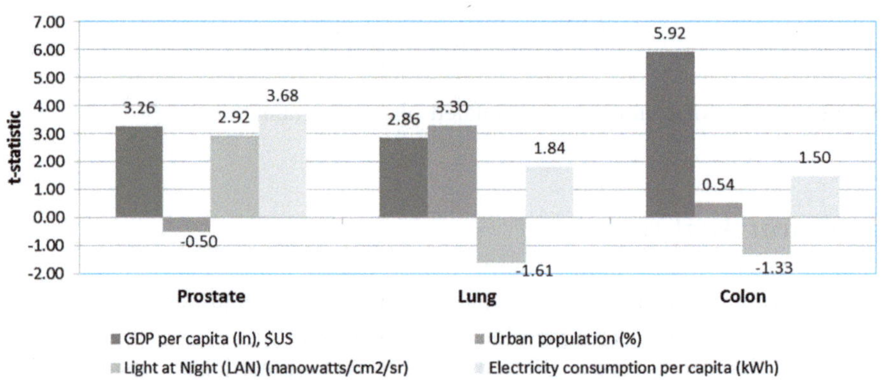

Fig. 13.2 *t*-statistics of selected variables affecting country-specific male cancers' rates (*Note* Full-fledged model specifications can be found in Kloog et al. 2009)

Table 13.1 Sensitivity test of PC ASR to plausible changes in the ground LAN intensity

LAN level	Average LAN value ($nW/cm^2/sr$)	Estimated ACR (per 100,000 residents)	Percent change
Low	8.60	66.77	–
Medium	28.95	87.11	30.5 %
High	99.21	157.01	80.2 %

Notes (1) The values of the fixed variables were fixed as follows: GDP *per capita* = $US 9,000 (the average value for the "high resource" countries under study); Urban population = 65.3 %, Electricity consumption *per capita* = 131.870 kWh. In addition, all regional indicators were set to 0 (that is, ASRs are estimated for the "high resource" (i.e., developed) countries)

Sensitivity Test

To estimate the relative contribution of LAN to PC ASRs, we split all the countries in our sample into three groups—countries with minimal LAN exposure (<15 $nW/cm^2/sr$), countries with average LAN exposure (15–57 $nW/cm^2/sr$), and countries with the highest LAN exposure (>57 $nW/cm^2/sr$). The Jenks "natural breaks" method of the ArcGIS™ software was used to classify countries into the groups. This method determines the best arrangement of values into classes by comparing the sum of squared differences of values from the means of their classes and thus identifies "break points" in the data values by picking the class breaks that best group similar values and maximize the differences between classes (Minami and ESRI 2000). Next, the values of all other variables (apart from LAN) were set constant to the average values observed in each group, and a sensitivity test of PC ASRs to changes in LAN values was run, using the multivariate model discussed in the previous subsection. The results of the sensitivity test are reported in Table 13.1.

As Table 13.1 shows, with the values of all other variables kept constant, the increase of LAN from 8.60 $nW/cm^2/sr$ (the average LAN value in the group of countries with minimal LAN exposure) to 28.95 $nW/cm^2/sr$ (countries with average LAN exposure) corresponds to an increase of 30.5 % in PC ASR. A further increase in LAN value to 99.21 (the maximum average LAN exposure) corresponds to an increase of 80.2 % in PC ASR.

Case Study 2: Ambient LAN and Breast Cancer in Women

BC is a leading cause of cancer death in women[2] (after lung cancer) and is the most common cancer among women worldwide, excluding non-melanoma skin

[2] The subsection is based on: Kloog I., Stevens R.G., Haim, A. and B.A. Portnov, Nighttime light level co-distributes with breast cancer incidence worldwide, *Cancer Causes & Control,* 2010, 21: 2059–68.

cancers. According to the World Health Organization (WHO), 1,300,000 women are diagnosed with BC annually and about 465,000 die from this disease every year (ACS 2007).

Parkin and colleagues (Parkin et al. 2001a, b, c; 2005) conducted several comprehensive studies of worldwide differences in cancer rates. During this research, the estimates of prevalence, mortality and incidence of 26 most common cancers were collected for 20 geographic regions of the world. The rates of BC in women appear to differ widely across the globe, being generally higher in developed countries than in developing ones. Thus, in North America age standardized rates (ASR) of BC incidence stand around 92.7 per 100,000, compared to Africa where BC ASR is 21.5 per 100,000 (see Fig. 12.1). Global variation of lung cancer is also large with ASR reaching 33.85 per 100,000 in North America *versus* 1.61 per 100,000 in Africa.

In the present subsection, we report results of the study which investigated the association between LAN and BC in women, using ASR of common cancers in women available for 164 world countries worldwide. If LAN is significantly associated with BC, as we hypothesized from the outset of the analysis, then an elevated incidence of BC with elevated levels of LAN can be expected, but not elevated risk of colorectal, larynx, liver and lung cancer, the cancers which are not hormone dependent and thus included in the present analysis as negative controls.

Cancer Data

As in our first case study (cancers in men), data on cancer ASR in women for the present analysis were obtained from the GLOBOCAN 2002 database, maintained by the IARC (Parkin et al. 2005). The IARC cancer data are reported for individual countries of the world for the period of 1998–2002. This data source has been previously used widely in epidemiological research (cf. e.g., Boyle and Ferlay 2005).

The data were obtained for breast, lung and colorectal cancers (three of the most common cancers in women) as well as for larynx cancer and liver cancer, the main risk factors for which are relatively well known (smoking for larynx cancer and hepatitis B virus (HBV) or hepatitis C virus (HCV) for liver cancer).

Explanatory Variables

Several development indicators of the world countries were included in the present analysis—population-weighted LAN, GDP per capita ($US), Percent urban population, Electricity consumption (kWh *per capita*)—as potential predictors of country-specific cancer incidence rates and perhaps as confounders of any possible LAN effect (see the discussion of the potential importance of these factors elsewhere in this chapter).

Fertility rate (average number of births per woman), was also included in the analysis as an additional explanatory variable for country-specific BC rates. As demonstrated by several empirical studies, fertility is negatively associated with BC risk (e.g., Kelsey and Gammon 1990). Fertility rates used in the analysis, to account for this effect, are total fertility rate (TFR), which is a more accurate measure of fertility than crude birth rates, since they refer to the average number of births per woman, rather than to average natural growth for population as a whole. In the present analysis, TFRs are used as an explanatory variable for BC only, since we have no a priori evidence that they may be associated with other cancer types under analysis.

Statistical Analysis

As in the previously discussed case study, the analysis was performed separately for each cancer type using the following linear model:

Cancer incidence rate = B0 (constant) + B1* (Electricity consumption) + B2*(GDP *per capita*) + B3*(LAN) + B4*(Percent of Urban population) + B5*(fertility rate) + ε (random error term), where B0, …,B5 are regression coefficients.

Factors Influencing BC Rates

Figure 13.3 reports the statistical significance of selected variables affecting female cancers' rates (as estimated by *t*-statistics from multivariate regression models). As in the previously discussed ("Cancers in Men") study, absolute values of *t*-statistic greater than 1.98 indicate a 0.05 (two-tailed) significance level, while *t*-statistic values above 2.61 indicate a 0.01 (two-tailed) significance level.

The models reported in Fig. 13.3 are estimated separately for the following five cancer types: breast, colorectal, larynx, liver and lung. Two regression models (1 and 2) are reported separately for BC incidence rates. These models differ in that Model 2 omits the five "outlier" Gulf States, which are present in Model 1.

The models for *breast, lung* and *colorectal* cancers provide good fit ($R^2 = 0.571$–0.648) and have a high degree of generality ($F = 41.975$–59.893, $P < 0.01$), while the *liver* and *larynx* models have poor fits (0.018–0.125), thus implying that predictors included in these models do *not* explain well the variability of these cancers across the globe.

Notably, among all the cancer types analyzed, only BC exhibited a significant positive association with *LAN exposure* (Breast (1): $t = 2.37$; $P < 0.05$; Breast (2): $t = 4.33$; $P < 0.01$). For all other cancer types LAN exposure was found to be statistically insignificant. *Per capita* GDP (ln) is also positively associated with ASRs of breast, lung and colorectal cancer ($P < 0.01$) while it is inversely

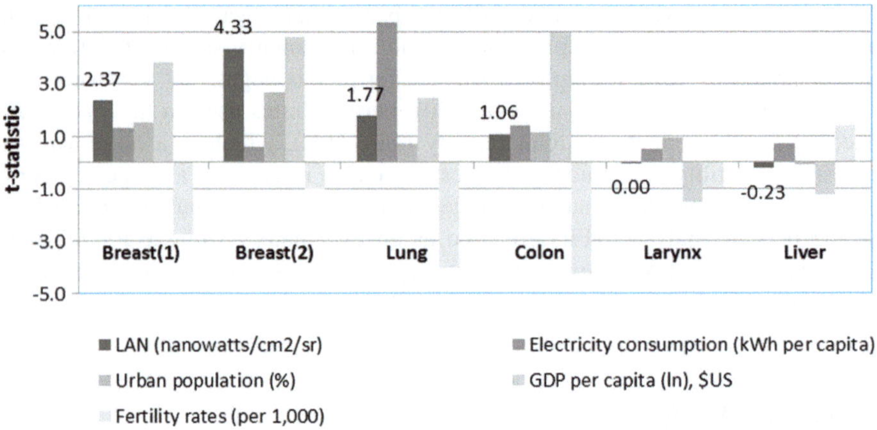

Fig. 13.3 Statistical significance (*t*-statistics) of selected variables affecting country-specific cancer rates in men (*Note* Full-fledged model specifications can be found in Kloog et al. 2010)

associated with liver cancer, albeit the association is not significant ($P > 0.05$). Fertility rates are negatively associated with BC as well as lung and colorectal cancer ($P < 0.01$), but not with larynx and liver cancer ($P > 0.3$), as could be expected.

Sensitivity Test

To estimate the relative contribution of LAN to BC calculated for ASRs, we split all the countries in our sample into three groups—countries with minimal LAN exposure (less than 15 nW/cm^2/sr); countries with medium LAN exposure (15–57 nW/cm^2/sr), and countries with the highest LAN exposure (greater than 57 nW/cm^2/sr). The Jenks "natural breaks" method was used to classify countries into the groups. Next, the values of all other variables from the second model (apart from LAN) were set constant to the average values observed in each group, and a sensitivity test of BC calculated for ASRs to changes in LAN values was run, using the BC models reported in Kloog et al. (2010). The results of the sensitivity test are reported in Table 13.2.

As Table 13.2 shows, when the values of all other variables are kept constant, the increase of LAN from 8.60 nW/cm^2/sr (the average LAN value in the group of countries with minimal LAN exposure) to 28.95 nW/cm^2/sr (countries with average LAN exposure) corresponds to an increase of 7.2 % in BC ASR. A further increase in LAN value to 99.21 (the maximum LAN exposure) corresponds to an increase of 23.25 % in BC ASR. There were five countries that had high fertility but also very high LAN exposure; these were all five Persian Gulf states (Saudi Arabia, Oman, United Arab Emirates, Qatar, and Kuwait). When these five

Table 13.2 Sensitivity test of breast cancer ASR to plausible changes in the ground LAN intensity

LAN level	Average LAN value ($nW/cm^2/sr$)	Estimated ASR (per 100,000 residents)	Percent change
	BC (1) Model (see Fig. 13.3)		
Low	8.60	40.47	–
Medium	28.95	43.39	7.20
High	99.21	53.43	23.25
	BC (2) Model (see Fig. 13.3)		
Low	8.60	44.45	–
Medium	28.95	50.08	12.70
High	99.21	69.54	38.85

Notes ASR-Age Standardized Rates per 100,000 residents. The values of the fixed variables were set constant as follows: GDP *per capita* = $US 9,000 (the average value for the "high resource" countries under study); Urban population = 65.3 %, Electricity consumption *per capita* = 131.870 kWh, fertility rate = 3.4 per 1,000 births

"outlier" Gulf States are omitted (Breast (2) model), the estimated BC as ASRs rises by about 50 % from the highest to the lowest LAN countries.

We also fitted the model to the 80 countries with a *per capita* GDP of >$3000 in order to partially control for a possible bias in the quality of the registries in the GLOBOCAN database. Parameter estimates were virtually unchanged compared with the full analysis of all 164 countries.

Interpretation of Results

There are considerable regional differences in PC ASRs in men and BC ASRs in women. While countries in Asia, Middle East, Africa, and Eastern Europe (often considered "low resource" areas) exhibit relatively low rates of these cancers, most developed countries exhibit very high cancer rates. These differences may be due to a variety of factors, including genetic background (explaining about 5 % of cancer incidence), economic status, differences in diets, amount of physical activities, obesity, exposure to environmental pollutants, and differences in medical care, among others.

As our analysis indicates, the relative risk of contracting cancer is positively associated with average income of local residents (Bray et al. 2004; ACS 2006). In part, but not in all, this excess is probably due to better access to medical and diagnostic procedures in the "high resource" societies (Wells and Horm 1992; Bradley et al. 2002; Madison et al. 2004). In addition, there appears to be a positive association between income, urban population, and ASRs of lung cancer. This association may be attributed to the fact that people living in urban areas are more exposed to air pollution emanating from industries and motor vehicles as well and to environmental (second-hand) smoking, due to high residential densities.

We found a significant positive association between *exposure to LAN, electricity consumption, and PC in men and BC in women*. There is of course the potential for confounding by known and unknown factors for which we could not adjust. However, for this to have occurred, the relative risk associated with the confounder would have to be very high, and the confounder would also have to be tightly correlated with LAN exposure in the localities studied (Blair et al. 2007).

The estimate of *per capita* LAN exposure was a novel aspect of the two case studies discussed in this chapter. This estimate was calculated as the average LAN exposure *per* person in each country under study. If there is no considerable misbalance between proportional shares of male and female populations in localities, the index in question provides a fairly unbiased LAN exposure estimate for both men and women.

Due to limitations on data availability, other risk factors, including occupation, alcohol consumption, smoking, or other factors could not be addressed by the analysis, though the *per capita* income variable may capture some of their effects. In addition, it should be noted that dynamics in population movement as well as behavioral patterns that limit exposure to LAN were also not covered by the study due to unavailable data. Such information can be obtained by studies carried out on a smaller scale, such as localities within an urban space, but not on a global level. Another potential limitation is in the completeness of cancer detection and registration in developing countries where LAN exposure is low. Parkin et al. (2001c) conducted a detailed analysis of cancer registration in Kampala, Uganda over the period 1994–1996 and concluded that "…it gives reassurance that published incidence rates are reasonably accurate."

In many cases, environmental health problems are better addressed by large-scale population level studies than by individual-level investigations, due to the occurrence of a large number of low-level exposures (Pekkanen and Pearce 2001). However, a substantial drawback of population-level studies is those associations that occur at an aggregated level may be subject to ecological confounding or fallacy (Robinson 1958; Selvin 1958). Several techniques were used to reduce the possibility of ecological confounding including grouping by geographic areas and adjusting for some potential confounders, such as income levels (GDP *per capita*) and percent urban population, which, among others, reflects population density (Morgenstern and Thomas 1993; Elliot 1996).

Finally, we should note that the associations between the DMSP measurements and BC&PC incidence discussed in this chapter *do not* necessarily lead to a conclusion that even very low light penetrating our homes from outside spaces, when we sleep, can increase BC&PC risks. However, people who live in communities which, according to Stevens (2011), "shine brightly to a satellite at night," are exposed to LAN not only in their bedrooms. The residents of such communities are also likely to be exposed to LAN in other places, and from a variety of other light sources, such as public transportation, shopping centers and various places of sports and entertainment as well as from advertising lighted advertisement boards, which residents of smaller towns and rural areas do not have, at least on such a scale. The circadian rhythms of the residents of such population hubs may also be disrupted by various nighttime activities, enabled by LAN, such as leisure and entertainment, and employment in businesses working after dusk. In this sense, nighttime satellite photometry, serving as a proxy for LAN-enabled daily rhythm disruption and a direct measure of ambient LAN pollution, helps to capture these additional LAN-associated risks (Haim and Portnov 2011).

Chapter 14
Selected Methodological Issues of LAN-BC&PC Research

Abstract Several methodological flaws may bias the results of a study of the LAN-BC&PC association. In this chapter we shall deal in brief with three such issues being in our view especially relevant to population-level studies—ecological fallacy, recall bias and eyelid effect.

Keywords Population-level studies · Methodological issues · LAN · Breast and prostate cancers (BC&PC) · Ecological fallacy · Recall bias · Eyelid effect · Bias preconditions · Potential remedies

> *If the doors of perception were cleansed everything would appear to man as it is, infinite.*
>
> William Blake

Ecological Fallacy

In social and environmental sciences, ecological fallacy is known as an incorrect assumption about individuals based on aggregate data for a group (Selvin 1958; Rothman et al. 2008). In other words, "ecological fallacy" (or "ecological bias") refers to differences in conclusions which may be drawn from group-level data, as opposed to data obtained for individuals.

Since population–level studies of the LAN-BC&PC association (such as, for instance, Kloog et al. 2008, 2009b, 2010) use country-level or regional estimates of LAN as surrogate measure for individual exposure, the following questions may arise:

> Do the exposure metrics, used in population-level studies of the LAN-BC&PC association, reflect, with sufficient accuracy, the LAN exposures of individuals and what solutions can minimize the risk of detecting spurious relationships attributed to ecological fallacy?

A. Haim and B. A. Portnov, *Light Pollution as a New Risk Factor for Human Breast and Prostate Cancers*, DOI: 10.1007/978-94-007-6220-6_14,

In his seminal paper, Robinson (1950) distinguished between two types of correlation—ecological and individual. The former is obtained for a group of people, while the latter is estimated for indivisible units, such as individuals. According to Robinson's line of argument, ecological and individual correlations tend to be dissimilar. As a result, any assumption about an individual based on average data obtained for a group to which the individual belongs may result in an assessment error, known as "ecological fallacy" or "ecological bias" (Selvin 1958; Rothman et al. 2008; Greenland and Morgenstern 1989; Elliott et al. 1992; Morgenstern and Thomas 1993).

The awareness about ecological fallacy has not affected geographic research at any considerable extent, where aggregate data are used widely for both empirical analysis and forecasting (see, for example, Glaeser et al. 1992; Felsenstein and Portnov 2005). However, in social and epidemiological studies, the situation is different. Due to the "ecological fallacy" concern, the use of aggregate data in these studies is being treated with caution (Openshaw 1984; Elliott and Wartenberg 2004; Greenland and Robins 1994; Greenland 2001).

Characteristically, the most striking example Robinson (1950) drew from the US 1930 Census of Population and Housing to substantiate his findings—illiteracy versus percent of foreign born,—had relatively little to do with "ecological fallacy" *per se*, but rather with the choice of input variables. While at the individual level, foreign immigrants in 1930 were generally less educated than "veteran" Americans, the aggregate data in Robinson's study (*ibid.*, p. 354) seemed to indicate otherwise: the correlation between percent illiterate (in a region's total population) and percent foreign born was found to be negative, implying that immigrants were more literate than the "natives." However, if average illiteracy rates were estimated for the foreign born (as opposed to the total population of regions, calculated by Robinson), the above spurious correlation between immigrant shares and illiteracy rates would have been avoided.

Sources of Ecological Bias

Greenland and Morgenstern (1989) single out two main Ecological Bias (EB) sources: omitted intra-group confounders, and effect modification. The former (i.e., omitted viable bias) refers to "the failure of a crude (or partially adjusted) association to properly reflect the magnitude of the exposure effect, due to differences in the distribution of extraneous risk factors among exposed and unexposed individuals" (*ibid.*, p. 269). As the authors of this review note, "there will be no ecological bias if both the background (unexposed) rate of disease and the exposure effect do not vary across groups, and there is no confounding within group." (*ibid.*, p. 273). Environmental smoking and nutrition deficiency are two examples of the most important ecological confounders which levels may vary across groups or regions, causing biased estimates of health effects attributed to exogenous environmental risk factors (Morgenstern and Thomas 1993).

However, as Portnov et al. (2007) demonstrate, ecological bias may arise even though none of the above conditions is violated. According to the results of this study, even though background health status and exposure effect are held constant, and no confounding at the individual level is present, the way in which the exposed and unexposed individuals are distributed across regional subdivisions may affect results. For instance, if the residents of each geographic region under study are *equally* affected by an environmental factor (that is, there is no significant variation in either interregional or intraregional exposure rates), then, indeed, the aggregation of data may lead to biased effect estimates. However, if substantially different levels of exposure are found among geographic regions, no ecological fallacy in data interpretation should presumably occur, and the linkages identified for areal aggregates are likely to emerge at the individual level as well (*ibid.*).

Ecological bias, analogical to confounding, may also occur when the background rate of a disease varies across unexposed populations or due to the presence of an "effect modifier" or a factor which is not necessarily a risk factor by itself (e.g., nutritional deficiency), but may modify the effect of the risk factor under study (e.g., smoking), due to a covariance between the two (Greenland and Morgenstern 1989; Greenland and Robins 1994) (*ibid.*, p.270).

Morgenstern and Thomas (1993) define two additional sources of ecological bias, viz. selection bias and information bias. While the former (i.e., selection bias) refers to the way in which research subjects are selected from the studied population, i.e., lost subjects or missing data. Concurrently, the latter type of ecological bias refers to "information loss" due to aggregation or measurement inaccuracy which may distort the effect estimates (Greenland 2001; Gotway and Young 2002).

Ecological bias may also occur due to non-linearity of relationships between individual risk factors. Hence the nonlinearity of relationship is not easily to detect in individual level data, it may lead to erroneous estimates, if a standard linear ecological model is used to approximate the relationships between covariates which actual association is non-linear and non-additive (Greenland and Robins 1994).

Recent epidemiological studies, specifically those using geographic information systems (GIS) technology (see *inter alia* Gotway and Young 2002; Elliott and Wartenberg 2004; Wakefield and Shaddick 2005; Bell 2006), highlighted another possible source of EB, known as "error propagation" (Goodchild et al. 1992; Heuvelink and Burrough 1989). The major cause of this sort of information bias emanates from the use of interpolation models converting data available for irregularly-spaced points (e.g., air monitoring stations) into regular grids or local exposure estimates at places were study subjects reside. Upon such interpolation, any error (e.g., attributed to faulty measurements), potentially occurring at original observation points, propagates into all output layers of data created by interpolation (Gotway and Young 2002; Bell 2006).

Furthermore, the outcome of interpolation can also be sensitive to the interpolation method used, i.e., nearest monitor, spline, inverse distance weighted (IDW), or kriging (McCoy and Johnston 2001). The choice of an interpolator and its specific quantitative parameters (such as power, search radius, etc.) may result in fairly different exposure estimates at the "target points" and, ultimately, in erroneous associations between exposure to an environmental risk factor and its health effects (Gotway and Young 2002; Elliott and Wartenberg 2004; Wakefield and Shaddick 2005).

Summing up, the following five criteria can thus be singled out as essential preconditions for minimizing the probability of EB in population-level studies:

- Securing the invariability of background rates of a disease and of exposure effect;
- Absence of significant confounding within the groups;
- Correct model specification accounting for a potential non-linearity of relationships between research variables;
- Minimizing the error propagation due to faulty measurements, missing data and/or errors attributed to interpolation, and,
- Detecting significant differences in both interregional and intraregional exposure rates prior to analysis.

Ways of Minimizing a Possibility of EB in Population Level Studies of the LAN–BC&PC Association

As we already mentioned elsewhere in this book, population-based and individual level studies have their own advantages and disadvantages. Global studies of large populations may provide a high degree of generality and thus help to capture the effects of low exposures by comparing a wide range of differently exposed subjects. However, population-level studies are generally considered weak in supporting causality (Kloog et al. 2010). If traditional case–control and cohort studies point one direction, but ecological studies point the other way, then there is a lack of coherence in the total body of evidence, which needs to exist at multiple resolution levels according to Hill's (1965) causality criteria (Stevens 2011).

Population-levels studies also often overlook detailed characteristics, such as hereditary factors, residential history and occupational risks. Yet, the results of such studies can be followed up by individual level studies in humans and by using animal models to understand mechanisms down to the cellular level (Haim et al. 2011).

The results of population studies are thus important in providing context for studies in subpopulations of people using case–control and cohort designs. If no association is revealed at the population level in a study with good statistical power, then that would be evidence against a strong effect of a putative risk factor. A significant association is thus a necessary, but not sufficient condition for a significant effect of a common exposure on risk in society at large (Kloog et al. 2010).

In Table 14.1 we suggest several empirical solutions aimed at minimizing a possibility of ecological bias (EB) in population-level studies of the LAN-BC&PC association, according to the EB criteria specified at the end of the previous subsection.

Table 14.1 Criteria for minimizing the possibility of EB in population-level studies of the LAN-BC&PC association and possible solutions

Criteria	Possible solutions
Securing the invariability of background rates of a disease and of exposure effects	Comparing disease rates across geographic units of the analysis (e.g., countries or regions) exhibiting comparable LAN levels: low variability of such rates will be indicative that the "invariability" criterion is not violated
Absence of significant confounding within groups	Accounting in the analysis for most known BC and/or PC risk factors, viz., age; ethnicity; reproductive history; oral contraceptive use; hormone therapy; alcohol consumption, obesity; physical activity; diet and vitamin intake; exposure to environmental chemicals; tobacco smoking, and nightshift working. Potential population-level proxies for factors which cannot be estimated directly at a general population level may include: population age structure (unless ASRs are used); local fertility rates; local *per capita* incomes (as proxies for obesity, physical activity, diet and vitamin intake); population densities and/or urbanization rates (as proxies for physical activity and smoking rates, environmental pollution and overall stress); percent of shift-workers (whenever available)
Correct model specifications accounting for non-linearity of relationships between research variables	Testing several functional forms (logistic, exponential etc.) with a subsequent comparison of model fits in respect to the variability explained
Minimizing error propagation due to faulty measurements, missing data and/or errors attributed to interpolation	Reliance on *in situ* measurements, rather than on self-reported exposure histories; comparing results of alternative interpolation techniques according to standard errors of the estimate, using GIS tools such as geo-statistical analysis
Detecting significant differences in both interregional and intraregional LAN exposure rates prior to analysis	Comparing intraregional and intraregional variations of LAN exposure and verifying that minimal variability exists (e.g., Mean ± SD)

Recall Bias

As Rafael (1987) suggests, any research which relies on questionnaires and self-assessments of past exposures may be a subject to the so called "recall bias" (RB). Such bias may occur when "cases" and "controls" recall their past exposures with different accuracy (*ibid.*). RB is thus a sort of more general "information bias" which is well known to researchers (Copeland et al. 1977).

The distinction should, however, be made between RB and simple "recall error." "Recall error" (or "non-differential recall") is an inaccurate assessment of past experiences or/and past exposures, attributed to age, education, and time period involved. In contrast, RB reflects significant differences between "cases" and "controls" in remembering past exposures and/or events.

According to Coughlin (1990), RB tends to be greater when recall is poor in general. In other words, "recall error" can simply be interpreted as a "memory failure", which may occur among both "cases" and "controls" but *not* necessarily leads to measurement errors, but only to a "loss of statistical power with a bias toward the null".

In contrast, consequences of RB, reflecting differential "memory failure" among "cases" and "controls" may be more severe, by effectively biasing "hypothesis tests away from the null" and potentially leading to incorrect effect estimates (Rafael 1987).

What Do We Know About RB and Its Potential Causes?

It should be acknowledged that empirical studies of RB are scarce, and, their estimates, in most part, are not always consistent. Some of these studies (mostly pertinent to cancer research) are summarized in Table 14.2 and discussed, in more detail, in the subsection below.

In an early survey, Stott (1958) reports a "classical" example of RB drawn from a post-WWII study carried out in Germany. In this study, 100 pregnancies resulting in major malformations were compared with 100 control pregnancies, with mothers, in both the case and control groups, being offered the same questionnaire after the birth. Emotional distress during pregnancies was mentioned by mothers of children with malformations more frequently than by mothers in the control group. However, a possibility that "emotional distress" *per se* was responsible for severe health effects reported in the study was ruled out by later studies, leading to a possibility that mothers of children with malformations tended to recall their "worries" and "emotional distresses" more often than mothers of the health born, simply because the former were apparently "looking harder" for feasible explanations for the negative birth outcomes (Geffeler 2008).

In another early study, Friedenreich et al. (1991) investigated a cohort of 325 BC cases and 628 matched controls that were offered the same self-administered questionnaire about dietary food habits, and distributed several years apart. The comparison of the results of both questionnaires, filled out by the same people but during different time periods, yielded little difference in the results, thus providing no evidence for any significant retrospective RB.

However, in another study of 491 children with acute lymphoblastic leukemia, compared with the same numbers of healthy population controls and the identical

Table 14.2 Summary of selected RB studies

Study	Research type or contingent	Methodology	Main findings
Friedenreich et al. 1991	A nested case–control study with 325 cases and matched 628 controls	Repeated self-administered questionnaire	Little difference was found in answering the questionnaires between "cases" and "controls", providing no evidence for a retrospective recall bias
Infante-Rivard and Jacques 2000	Children diagnosed with leukemia versus healthy children and hospital controls	Telephone interviews with children's parents carried out several years apart	Parents of children with leukemia neither provided more correct answers about location of power lines, than parents of healthy children and parents of hospital controls nor affected by leukemia
De Vries et al. 2005	Case control study of melanoma patients and population based controls	Calculation of crude odds ratios	Protective effect was found between sun exposure and melanoma risk and between sun exposure and naevus count recall
Boffetta et al. 2008	Critical survey	Critical survey and meta- analysis of published data	On the general, disease patients may have a greater motivation to explain their disease and thus are more likely to recall past exposures than healthy individuals
Vrijheid et al. 2009	Mobile phone records of breast tumor patients and healthy controls collected over a two year period	The ratio of reported to recorded phone use was analyzed as the measure of agreement	Mean ratios were found to be nearly identical for cases and controls, with both groups underestimated the number of calls and overestimated call duration
Parr et al. 2009	Melanoma patients and population-drawn controls with similar rates of response	Comparison of self-reported melanoma risk factors	Shifts in responses were observed among both cases and controls, but a shift in cases was observed only for skin color after chronic sun exposure, and a larger shift in cases was observed for nevi

number of hospital controls with severe hematologic diseases, carried out by Infante-Rivard and Jacques (2000), RB was found to be present. In particular, the research revealed that, in recalling the distance to power lines from their homes, the parents of children with leukemia provided 62 % correct answers, as opposed to only 22 % correct answers among population controls and 36 % among hospital controls.

de Vries et al.'s (2005) study of 597 melanoma cases and 622 population based controls also found some indirect evidence about potential RB. In this study, a negative association was found between sun exposure and melanoma risk (OR = 0.87; 95 % CI: 0.65–1.18) and between sun exposure and naevus count recall, raising a possibility that cases may have underreported their sun exposure, potentially resulting in RB.

In another study, Fransson et al. (2008) compared answers recorded in 2005 to those recorded in 1992–1998 (regarding past leisure time, occupational, and physical activity levels) by 78 patients with a myocardial infarction and 243 control subjects in Sweden, also found some (albeit statistically weak) support for RB presence. In particular, people who recalled the same activity level as originally reported was nearly identical ranging from 69 to 96 % among cases and from 69 to 89 % among controls. However, the perceived physical workload in the household was found to match retrospective records better among controls, indicating the presence of potential RB in recalling past household physical activity levels.

Yet, in a separate study, Vrijheid et al. (2009) analyzed mobile phone records of 212 cases and 296 controls collected from network operators over a 2 year period, and compared these actual data with mobile phone use reported during an interview. The ratio of reported to recorded phone use was used in the study as the measure of agreement. Mean ratios were found to be nearly identical for both cases and controls. Both groups underestimated number of calls by a factor of 0.81 and overestimated call duration by a factor of 1.4. However, for cases, but not controls, ratios increased with increasing time before the interview, leading to a possibility that RB may be present in the data, especially regarding earlier time periods.

Essential Preconditions for RB

Based on the above studies of the RB phenomenon, four possible preconditions for its occurrence can thus be singled out:

- *Prior knowledge of a potential health risk.* People affected by a disease are more likely to remember and/or report past exposures to certain factors (which harmful effects they may be aware of) than are healthy individuals, creating difference in assessment which may result in over-reporting of past exposures by research subjects, compared with controls;

- *Guilt or remorse*. People unaffected by a disease may be less likely to recall a true exposure to a certain risk factor to which they were exposed voluntarily (e.g., drug use or alcohol consumption during pregnancy);
- *Debilitating effect of a disease on the memory function*. A disease may have a debilitating effect on human memory (e.g., causing memory loss), preventing people affected by it from recalling accurate actual past events and exposures;
- *Memory substitution*. Live styles and daily habits of a person can change after a disease's diagnosis. As a result, questions about "exposures in general" may not accurately represent past exposures in the "case" group, that is, exposures which are more relevant for risk assessment, due to a latency period, than present-day exposures.

Likelihood of RB in LAN-BC&PC Studies

Considering these conditions, should RB become a point of concern for studies of the LAN-BC&PC association?

In several publications, such a possibility had indeed been raised (cf. e.g., Davis et al. 2001; Shernhammer and Stone 2011).

The first three abovementioned RB criteria can, apparently, be dismissed outright in the case of the LAN-BC&PC association, on several grounds. First, we can confidently assume that most people are rather unaware of potentially harmful health effects of LAN and are thus not expected to "overestimate" this potential risk factor. Despite several publications in mass media on the topic, these studies have not yet generated, in our view, a wide-range concern about LAN effects among the general public, as it happened with e.g., microwave radiation, smoking, and cellular phone use.

Second, people are unlikely to feel any remorse about LAN exposure (apart from costs involved) and nighttime activities enabled by LAN. We can thus safely reject the second (that is "guilt and remorse") criterion as well. BC&PC are not known to have a "debilitating" effect on human memory either.

However, the fourth RB criterion—i.e., memory substitution—should not be taken lightly. As Davis et al. (2001), justly noted, "it is possible that a woman's recall of prior sleep habits could be affected by her more recent disease experience, resulting in differential recall for case patients relative to control subjects." However, the authors of this paper also remarked that such an effect is highly unlikely if questions in a LAN-related questionnaire are professionally formulated, for instance, if the subjects are not asked about "restlessness or sleeplessness" but about sleeping or not sleeping during particular hours during nighttime.

Optimally, LAN exposure metrics should be estimated by *actual measurements*, as it was done, for instance, in Rea et al. (2011) study, rather than by questionnaires. However, such an approach would not always be practicable when hundreds or thousands of subjects and controls are involved. *In situ* measurements cannot account for past exposure either.

In the absence of *in situ* measurements, the researcher can formulate LAN related questions twice: first related to present day exposures and, for the second time, elsewhere in the questionnaire, as related to past exposures, which occurred before the disease's diagnosis. Questions about life-style variables and daily habits before and after the disease's detection can also be included, so as to indicate the magnitude of changes and estimate the strength of potential RB.

Lastly, we should note that we are unaware of any study that investigated the presence of RB in relationship to LAN exposure. Only such studies, if carried out in the future, will tell us with certainty whether self-reported LAN exposures are consistent with actual LAN levels, but, more importantly, whether such recalls do differ between "cases" and "controls" in recalling bedroom and outdoor light levels, which is a necessary precondition for RB to occur.

Eyelid Effect as an Additional Potential Source of LAN Exposure Bias

Most people sleep with their eyelids closed. Since, according to one of the proposed mechanisms of the LAN-BC&PC association, the direct exposure of human eyes to LAN tends to reduce pineal MLT secretion, the question thus becomes: *Is LAN capable of penetrating eyelids during the sleep time, even if eyelids are tightly closed*?

There is evidence that bright light can lower MLT in some persons even through closed eyelids (Hätönen et al. 1999). However, in another study, no reduction of MLT secession was found by exposure to light when eyelids were kept closed (Jean-Louis et al. 2000). Yet, Czeisler et al. (1995) reported that plasma MLT concentrations decreased during exposure to bright light in blind patients by an average of 69 %. However, when blind patients were tested with their eyes covered during their exposure to light, plasma MLT did not decrease.

In interpreting these results it is important to keep in mind several considerations. The eyelids are not closed immediately after a person goes to sleep, and nor do they remain tightly closed throughout the entire nighttime period (Haim et al. 2011). Exposure to bright light even in people, sleeping with closed eyelids may raise body core temperature (thus meaning that pineal MLT production is suppressed), increase the "latency to sleep onset" and shorten the normal sleep period (Dijk et al. 1987, 1991).

It should also be kept in mind that most people are awake for brief periods during the night, which is normal and healthy (Wehr 1992), and sometimes one awakens to empty the bladder, and with increasing frequency with each passing year (Haim et al. 2011). Moreover, as we already mentioned elsewhere in this book, LAN is often characterized by a short wavelength (that is, emitted as green and blue lights), which is especially effective in pineal MLT suppression.

Another evidence about a possibility that eyelids are not "impassible barriers to LAN" comes from a recent study by Kloog et al. (2010), which indicated that bedroom light levels were significantly associated with BC risk ($P < 0.01$).

It should not also be overlooked that population studies normally deal with a "chronic exposure" to LAN, that is exposure which occurs for several hours each night, for many nights, and not with an "acute" exposure to relatively short light "episodes," which are usually used in laboratory experiments. Such exposures effectively minimize the "eyelid effect" by accounting for accumulated light-sleep interferences. As a result, we should not, in our view, be overly concerned about this sort of bias in population-level LAN-BC&PC studies.

Finally, we should emphasize that the direct exposure of eye retina to LAN when we sleep and the resulting suppression of MLT production is only one possible venue of LAN effect. Other possible venues, discussed elsewhere in this book, are daily rhythm disruption, and the role of LAN as a general stressor. Since most nighttime activities, enabled by LAN (such as working at brightly lit environments, TV watching, attending places of entertainment after natural dusk, etc.), are performed with eyelids open, the latter have little effect on preventing the adverse effects of LAN exposure and associated BC&PC risks.

Chapter 15
Dark-Less World: What is Next? (Conclusions and Prospects for Future Research)

Keywords Conclusions · Sustainable illumination policy · Health effects · Behavioral changes · Future research

> *A scientific truth does not triumph by convincing its opponents and making them see the light, but rather because its opponents eventually die and a new generation grows up that is familiar with it.*
>
> Max Planck

In past decades, humans have effectively defeated the "nighttime darkness" by "chasing it away" from our homes, streets, highways and public-spaces, both in western countries and in many countries of the developing world. LAN has spread rapidly due to relatively easy accesses to electricity and rapidly developing infrastructures, and seems to have penetrated even the most remote corners of our Planet. Indeed, nighttime illumination today is present not only in urban areas, but also in rural communities. In addition, we commonly use today, what we call the "environmental friendly illumination," which converts electrical energy more efficiently into short wavelength (bluer) illumination replacing more traditional incandescence bulbs, which were closer, in their spectral properties, to natural daylight.

Beyond doubts, the proliferation of artificial illumination has changed our life dramatically. Thanks to it we can work after the natural dusk, and extend our free time available for entertainment, sports and leisure. This is a welcome change in our lives, but, as we saw elsewhere in this book, we may well pay a high price for this change. *So what should we do in order to mitigate potential health risks associated with rapid LAN proliferation?*

Today, we pay much attention to global climate change, although even in the past, our Planet had gone through cycles of warmer and cooler periods. Yet, we do not pay enough attention to potential consequences of the change that LAN introduced into our lives, although never before (that is, beyond the most recent 130–140 year period) our Planet and its inhabitants had been exposed to LAN of such intensities and wavelength throughout the year.

If, in the relatively recent past, shift-workers were a special population group almost solely exposed to LAN, the situation today has changed dramatically. When we attend places of entertainment late at night or seat in front of our computers communicating with our friends over the Internet after midnight or

A. Haim and B. A. Portnov, *Light Pollution as a New Risk Factor for Human Breast and Prostate Cancers*, DOI: 10.1007/978-94-007-6220-6_15,

write an assignment for tomorrow's class, we become, even without realizing it, "shift-workers", suffering, from the same problems that shift-workers suffer from, namely, MLT suppression due to short wavelength LAN exposure and daily rhythm disruption, with all potential health risks such an exposure may entail.

Humans, in developed and less developed countries alike, are consumers of natural resources, one of which is electrical energy converted into illumination. Increasing electricity production is beneficial for economic development and, presumably, for societal security, but it also brings negative consequences for the environment, by increasing the CO_2 emission, which has become a major concern over past decades. Making artificial illumination more efficient by introducing energy-saving light sources, such as LED and florescent light tubes, helps to reduce CO_2 emission.

However, by introducing these light sources into our lives, we should be aware of potential negative impacts they may have on our health. Today if we put together chronobiological evidence about negative effects of short wavelength illumination and suppressing of MLT production at night, we may regret this step due to health hazards as well as economic and social costs associated with it, and, therefore, it should be considered unsustainable.

However, as we do not want to return to pre-Edison days of "nighttime darkness", the following question should be answered: *What should we do in order to enjoy LAN while reducing its negative effects on our health and wellbeing?*

The general knowledge and scientific evidence accumulated in past decades, and summarized in this book, may help to find some relatively simple answers to this question. First and foremost, we need to keep in mind *how* LAN may affect our health. Such potential causality pathways are featured in Fig. 15.1 and discussed, in more detail, in separate chapters of this book. Let us summarize them in brief. According to the rapidly increasing body of scientific evidence, LAN may reduce pineal MLT production levels directly by reaching the non-image forming photoreceptors of our retina. In Fig. 15.1 this "simplest" pathway is marked as the "LAN-MLT path."

Due to nighttime activities which LAN enables, it may also affect our health indirectly, by introducing behavioral changes. By skewing our daily schedule towards nighttime activities, we effectively alter our "diurnal nature" established by thousands of years of evolution and thus interfere with the normal functioning of our biological clock entrained by exogenous photic changes. In Fig. 15.1, this indirect pathway is marked as the "LAN-DRD [Daily Rhythm Disruption] path." According to scientific evidence provided by controlled laboratory experiments (Monteleone et al. 1992; Zisapel et al. 2005), physical activities during nighttime, even carried out in complete darkness, tend to reduce pineal MLT production.

Furthermore, since light is not anticipated during the dark phase, LAN may play a role of a general stressor. The stress response induces processes which are aimed to maintain homeostasis and involves increased levels of stress hormones and energy expenditure. As additional energy is allocated to maintain homeostasis, it may weaken the response of the immune system thus, increasing vulnerability to pathogens and carcinogens (the "LAN-SR [Stress Response] path"; see Fig. 15.1).

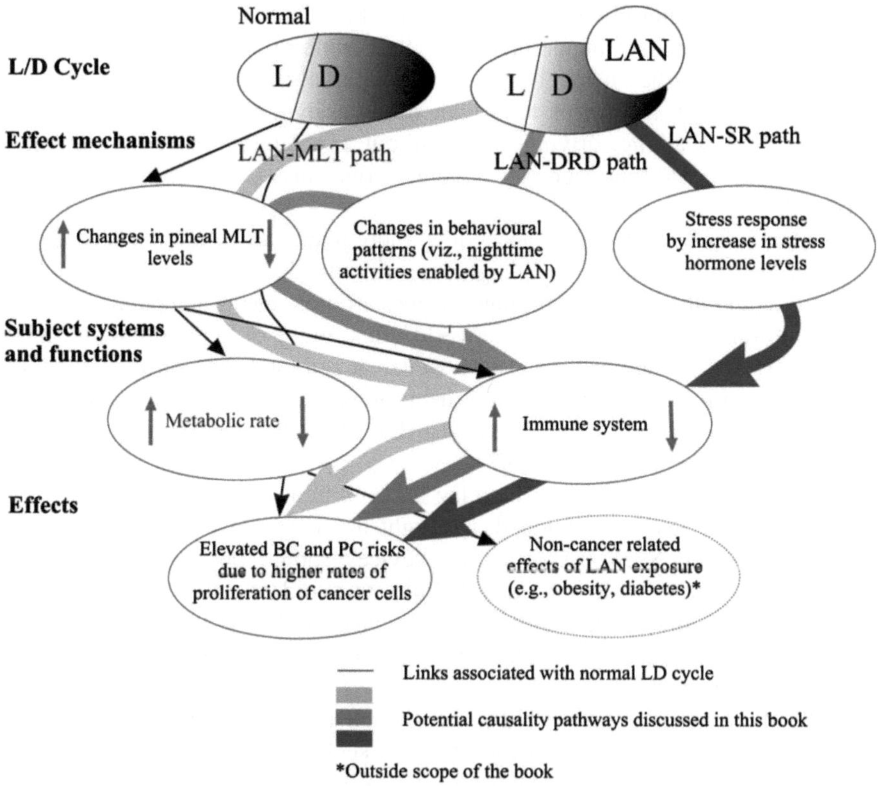

Fig. 15.1 Potential effects of LAN and causality pathways discussed in the book

Keeping this in mind, we should thus try to adhere to the "traditional" 12L:12D cycle, by avoiding, whenever possible, nighttime activities, including TV watching and computer gaming, especially for children. Considering that not too many of us have to work nightshifts, for the majority of us, giving preference to late night activities over daytime ones becomes the matter of choice, and we should be aware of potential consequences.

Second, when we rest, bright lights in our bedrooms should be avoided, both lights coming from indoor and outdoor light sources. Low intensity (yellow and red non-LED) lights may, in most cases, satisfy our needs in the nighttime illumination, mainly for the elderly, to find their way to the bathroom at night, and for children, to feel secured.

Preventing LAN penetration from outdoor sources (such as e.g., street lights, illumination of public buildings, and lights from moving vehicles) may also be relatively easy in most temperate climates, in which constant cross ventilation of indoor spaces by keeping windows open at night is not as necessary as in tropical climates. Even if we want to keep windows open during nighttime, the penetration of outdoor LAN can be reduced by closing curtains and shades.

Third, local authorities may contribute to the reduction of light pollution by implementing more sustainable nighttime illumination policies, such as, lowering illumination intensities in public places to minimally acceptable levels, by surveying actual needs and caring to provide minimal light levels required by certain activities (e.g., walking, driving, etc.) without over-illumination. In addition to health benefits, additional energy savings from such policy can naturally be expected.

As electricity becomes more accessible, more places are illuminated, especially (but not only) in urban areas. Nighttime illumination is commonly used for advertising and illuminating buildings, such as art monuments and public buildings being cultural landmarks all over the world. This source of light pollution, cannot, in our view, be avoided completely. However, it can (and should) be restricted to certain hours in the early dark period, using long wavelength illumination which has a minimal effect on reducing pineal MLT production, where after which the illumination should be switched off.

As we developed devices for tracking air pollutions, devices for tracking light pollution should also be developed, and light intensity above a certain threshold should be dimed. Such LAN-tracking devices should be able to distinguish between different wave lengths thus helping to avoid an extensive use of short wavelength illumination. In addition, street lights should not be directed at homes but rather be focused on pavements and road surfaces, in order to decrease light pollution in sleeping habitats.

Regulations on nighttime advertising (which is virtually non-existent today) should also be implemented and switching off such illumination can contribute to the most important aims of decreasing light pollution and energy-saving. Unfortunately, we are used to the fact that huge areas are lit at night without any actual need. This attitude should change and the light needs to be pointed at the objects it is meant to illuminate. For example, lampposts of 15 m above the road surface become a strong source of light pollution due to higher light intensity such lampposts emit.

Public policy makers should also be convinced to curb the wide scale introduction of short wave length illumination, which is potentially the most dangerous component of light pollution.

Education of consumers may also play a role, helping them to make informed choices. As we mark today alcohol and tobacco products, which have negative effects on our health that we are well aware of, we should also consider marking packages of short wavelength light bulbs with a consumer warning, such as, for instance: “This bulb emits light of short wavelength of less than 530 nm, and such illumination effectively suppresses pineal MLT production, when used at nighttime, and may thus endanger your health.” After reading this warning we can decide whether we want to buy this product or not. Although such a consumer warning may sound “overreacting” at this point of time, it may eventually come into practice, as the authors of this book believe, as research develops and new evidence about light pollution as a source of toxicity and its adverse health effects becomes available to the public.

More information about negative consequences of suppressing pineal MLT at night can be obtained from the internet and scientific publication, including this book, helping to learn about negative consequences of suppressing pineal MLT at night time.

Unfortunately, we are still lacking basic research in regards to light pollution and its various impacts on human and animal health on the one hand and stability of natural ecosystems on the other, albeit research from the epidemiological level down to the cellular level has been carried out by different research teams on BC&PC and their links to light pollution. From the literature we have strong evidence that in BC patients MLT levels are low. Bearing in mind BC latency period is of about 10 years, people with low levels of MLT which are not at the stage of cancer patients can be treated with MLT to prevent the BC or PC development. Therefore, the question is *whether earlier tests of MLT levels can be used as an indicator for a higher risk of developing BC or PC?*

A common technology for detecting BC these days is the use of mammography scanning. However, a positive result means the tested woman is already a BC patient. If we think of the results of our 2011 study in which we can assume that at list 40 % of BC cases are due to exposure to LAN in their sleeping habitat, many cases of BC can be prevented, if MLT levels are timely tested. If this hypothesis is correct, it will definitely help to formulate informed health policy, and the amount of funds devoted to BC treatment can be reduced. Scientific evidence accumulated to date and summarized in this book definitely lead us to believe that with right measures in regards to light pollution we can significantly reduce the incidence of BC and PC.

Unpublished results from the Israeli Center for Interdisciplinary Research in Chronobiology at the University of Haifa (Schwimmer et al. -in preparation) revealed in DNA extracted from mice BC cells, where the mice were exposed for 30 min each night to light (about 470 nm, and at an intensity of 450 lux), a loss of methyl groups (hypo-methylation) were noted when testing DNA global methylation. However, mice treated with MLT showed similar levels to those of the control group that was not interfered with LAN. Therefore, a combination of MLT levels (measured from salvia together with levels of DNA global methylation from white blood cells for instance can be used as biomarkers. MLT treatment and avoidance of exposure to short wave length illumination by using appropriate illumination can be used as means for prevention. However, these ideas should be transferred to medical protocols after wide scale surveys and experiments.

Blue wavelength blocker glasses that help to avoid MLT suppression (Sasseville et al. 2006) may also be a reasonable solution for home use and for shift workers who have sleep breaks during their night shifts. Of course, it is better to focus on sustainable illumination than to produce unsustainable illumination and then look for ways to overcome it. However, such glasses and other gadgets can be a temporary solution until more sustainable illumination are implemented, just like swallowing MLT pills before going to sleep at night.

As we have discussed elsewhere in this book, LAN can be a general stressor. To avoid its negative effect, on/off switches should be replaced by dimmers, which can increase light intensity gradually, in the way it happens under natural conditions, for instance, during sunrise, and darken the room, like at the sunset when we are indoors. By not moving to the abrupt darkness at once, we can adjust our vision system to the changing illumination.

We should also not to ignore the fact that the circadian system in mammals starts developing in the late stages of the embryonic development but begins to function only after birth. This actually means that the only MLT source at night for the developing embryo is his mother's pineal gland. Bearing in mind the important role of MLT in the cell function as an anti-oxidant thus raises a question about what happens if MLT in the mother is suppressed due to her LAN exposure. *Can such MLT suppression explain, at least in part, an increase in the number of health disorders in the young generation we do not know how to explain today?*

LED illumination (for instance, stand-by and control lights in various devises and equipment) is considered to be most energy saving. However, its introduction into our homes in so many forms should worry us the most. Although each one of them maybe of low intensity but when they are lit together, they may suppress nighttime MLT production. These devices should therefore be removed from our sleeping habitat and kept in a separate room. This is a simple solution that can reduce our exposure to LAN.

Some 50 or even 40 years ago measures which we implement today to minimize direct health damage from air pollution or cigarette smoking would have seemed unrealistic. But the situation has changed drastically in the past 50 years and we uphold today strict regulations and laws helping to avoid the damage caused by cigarette smoking and passive exposure to smoking and thus managed to decrease lung cancer incidence rates drastically in most western societies. *Why should not we try to do the same with light pollution?* Economic considerations are very strongly involved in the tobacco industry, as well as sale taxes, but social and environmental issues cannot be underestimated either.

The future illumination should be based on wave length, intensity, duration and frequency of exposure that do not put our health at risk and can be even used during night time in bedrooms mainly for the elderly and children. No doubt that the resolution recently passed by the American Medical Association (AMA) is an important step in the right direction. Even more medical organizations around the world should adopt such resolutions. If it happens, it will be much easier for decision makers to embrace the idea of sustainable illumination and act in an appropriate way.

Considering LAN as a source of environmental pollution and a source of toxicity is a challenge to our way of thinking. As we stated already in this chapter, we do not want to live in the dark "pre-Edison" world. Therefore, we need to implement dramatic changes in thinking paradigms. Realizing the negative effects of LAN should help decision-makers to support the development of a holistic approach helping to deal with the problem at large. Therefore, environmental considerations, together with economic and sociological factors, should be considered before introducing new technologies, which, at first glance, may sound environmentally friendly, but, at the end of the road, may be proved hazardous to the environment and to human health, with a wide range of negative socio-economic implications such hazards may entail.

References

ACS (2007) Global cancer facts and figures. http://www.cancer.org. Accessed 15 Feb 2008
Aggelopoulos NC, Meissl H (2000) Responses of neurones of the rat suprachiasmatic nucleus to retinal illumination under photopic and scotopic conditions. J Physiol 523:211–222
Akasawa T, Masuda H, Saeki Y, Matsumato M, Takeda K, Tsujimura K, Kuzu-shima K, Takahashi T, Azuma I et al (2004) Adjuvant-mediated tumor regression and tumor-specific cytotoxic response are impaired in MyDos-deficient mice. Cancer Res 64:757–764
Akbulut H, Icli F, Buyukcelik A, Akbulut KG, Demirci S (1999) The role of granulocyte–macrophage- colony stimulating factor, cortisol, and melatonin in the regulation of the circadian rhythms of peripheral blood cells in healthy volunteers and patients with breast cancer. J Pineal Res 26:1–8
Albrecht U, Zheng B, Larkin D et al (2001) mPer1 and mPer2 are essential for normal resetting of the circadian clock. J Biol Rhythms 16:100–104
Allen GC, West JR, Chen WJA, Earnest DJ (2005) Neonatal alcohol exposure permanently disrupts the circadian properties and photic entrainment of the activity rhythm in adult rats. Alcohol Clin Exp Res 29:1845–1852
Althuis MD, Dozier JM, Anderson WF, Devesa SS, Brinton LA (2005) Global trends in breast cancer incidence and mortality 1973–1997. Int J Epidemiol 34:405–412
Anisimov VN, Popovich IG, Zabezhinski MA (1997) Melatonin and colon carcinogenesis: I. Inhibitory effect of melatonin on development of intestinal tumors induced by 1,2-dimethylhydrazine in rats. Carcinogenesis 18:1549–1553
Anisimov VN (2006) Light pollution, reproductive function and cancer risk. Neuro Endocrinol Lett 27:35–52
Anselin L, Syabri I, Kho Y (2005) GeoDa: an introduction to spatial data analysis. Geogr Anal 38:5–22
Anselin L (1999) Spatial econometrics. Bruton Center, School of Social Sciences, University of Texas at Dallas, Texas
Armstrong SM (1989) Melatonin: the internal Zeitgeber of mammals? Pineal Res Rev 7:157–202
Aschoff J (1960) Exogenous and endogenous components in circadian rhythms. Cold Spring Harbor Symposium, Quant Biol 25:11–28
Aschoff J (1965) Circadian rhythms in man: a self-sustained oscillator with an inherent frequency underlies human 24-h periodicity. Science 148:1427–1432
Aschoff J (1990) From temperature regulation to rhythm research. Chronobiol Int 7:179–186
Aschoff J, Pohl H (1978) Phase relations between a circadian rhythm and its zeitgeber within the range of entrainment. Naturwissenchaften 65:80–84

A. Haim and B. A. Portnov, *Light Pollution as a New Risk Factor for Human Breast and Prostate Cancers*, DOI: 10.1007/978-94-007-6220-6,

Ashkenazi L, Haim A (2012) Light interference as a possible stressor altering HSP70 and its gene expression levels in brain and hepatic tissues of golden spiny mice. J Exp Biol 215:4034–4040

Aujard F, Herzog ED, Block GD (2001) Circadian rhythms in firing rate of individual suprachiasmatic nucleus neurons from adult and middle-aged mice. Neuroscience 106:255–261

Banerjee S, Wall MM, Carlin BP (2003) Frailty modeling for spatially correlated survival data, with application to infant mortality in Minnesota. Biostatistics 4:123–142

Banin D, Haim A, Arad Z (1994) Metabolism and thermoregulation in the Levant vole Microtus quentheri: the role of photoperiodicity. J Therm Biol 19:55–62

Baugh K, Elvidge C, Ghosh T, Ziskin D (2010) Development of a 2009 stable lights product using DMSP-OLS data. Proceedings of the 30th Asia-Pacific advanced network meeting, Hanoi, pp 114–130

Bell ML (2006) The use of ambient air quality modeling to estimate individual and population exposure for human health research: a case study of ozone in the Northern Georgia Region of the United States. Environ Int 32(5):586–593

Ben Shlomo R, Kyriacou CP (2010) Light pulses administrated during the circadian dark phase alter expression of cell cycle associated transcripts in mouse brain. Cancer Genet Cytogenet 197:65–70

Benenson W, Harris WJ, Stocker H, Lutz H (2002) Handbook of physics. Springer, New York, p 336

Berson DM (2007) Phototransduction in ganglion-cell photoreceptors. Pflugers Arch 454(5): 849–55

Blair A, Stewart P, Lubin JH, Forastiere F (2007) Methodological issues regarding confounding and exposure misclassification in epidemiological studies of occupational exposures. Am J Ind Med 50:199–207

Blask DE, Brainard GC, Dauchy RT, Hanifin JP, Davidson LK, Krause JA, Sauer LA, Rivera-Bermudez MA, Dubocovich ML, Jasser SA, Lynch DT, Rollag MD, Zalatan F (2005) Melatonin-depleted blood from premenopausal women exposed to light at night stimulates growth of human breast cancer xenografts in nude rats. Cancer Res 65(23):11174–11184

Boffetta P, McLaughlin JK, La Vecchia C, Tarone RE, Lipworth L, Blot WJ (2008) False-positive results in cancer epidemiology: a plea for epistemological modesty. J Nat Cancer Inst 100:988–995

Bonner MR, Nie J, Han D, Vena JE, Rogerson P, Muti P, Trevisan M, Edge SE, Freudenheim JL (2005) Secondhand smoke exposure in early life and the risk of breast cancer among never smokers (United States). Cancer Causes Control 16:683–689

Borisenkov MF, Anisimov VN (2011) Cancer risk in women: a possible connection with geographic and certain economic and social factors. Vopr Onkol 57(3):343–354. (in Russian)

Boyle P, Ferlay J (2005) Cancer incidence and mortality in Europe, 2004. Ann Oncol 16:481

Bradley CJ, Given CW, Roberts C (2002) Race, socioeconomic status, and breast cancer treatment and survival. J Natl Cancer Inst 94:490–496

Brainard GC, Hanifin JP, Greeson JM, Byrne B, Glickman G, Gerner E, MD Rollag (2001) Action spectrum for melatonin regulation in humans: evidence for a novel circadian photoreceptor. J Neurosci 21:6405–6412

Brainard GC, Rollag MD, Hanifin JP (1997) Photic regulation of melatonin in humans: ocular and neural signal transduction. J Biol Rhythms 12:537–546

Brandes O, Haim A, Zisapel N (2004) The effect of photoperiod on the reproductive system and melatonin secretion in the social vole (*Microtus socialis*). In: Feare CJ, Cowan DP (eds) Advances in Vertebrate Pest Management III . Furth: Filander Verlag, pp 49–60

Bray F, McCarron P, Parkin DM (2004) The changing global patterns of female breast cancer incidence and mortality. Breast Cancer Res 6:229–239

Brody JG, Moysich KB, Humblet O, Attfield KR, Beehler GP, Rudel RA (2007) Environmental pollutants and breast cancer: epidemiologic studies. Cancer 109:2667–2711

Bronson FH (1989) Mammalian reproductive biology. University of Chicago Press, Chicago

Bruning A, Holker F, Wolter C (2011) Artificial light at night: implications for early life stages development in four temperate freshwater fish species. Aquat Sci 73:143–152
Brzezinski A (1997) Melatonin in humans. New Engl J Med 336:186–195
Brzezinski A, Vangel MG, Wurtman RJ, Norrie G, Zhdanova I, Ben-Shushan A, Ford I (2005) Effects of exogenous melatonin on sleep: a meta-analysis. Sleep Med Rev 9:41–50
Bureau International des Poids et Mesures (2007) SI Brochure Appendix 2. Practical Realization of the Definition of the Candela
Cajochen C, Munch M, Kobialka S, Krauchi K, Steiner R, Oelhafen P, Orgul S, Wirz-Justice A (2005) High sensitivity of human melatonin, alertness, thermoregulation, and heart rate to short wavelength light. J Clin Endocrinol Met 90:1311–1316
Cerutti P, Ghosh R, Oya Y, Amstad P (1994) The role of the cellular antioxidant defense in oxidant carcinogenesis. Environ Health Perspect 102(Suppl 10):123–129
CIA (2006) CIA World Factbook. http://www.cia.gov/index.html. Accessed 2006
CIE (1926) Commission Internationale de l'Eclairage Proceedings 1924. Cambridge University Press, Cambridge
CIE (1931) Sec 4; Vol 3. In: Commission Internationale de l'Eclairage Proceedings, vol 1. Bureau Central de la CIE, Paris, p 37
CIE (2001) The CIE system of physical photometry draft standard 010.2/E
Cinzano P, Falchi F, Elvidge CD (2001) The first world atlas of the artificial night sky brightness. Mon Not R Astron Soc 328:689–707
Cinzano P, Falchi F, Elvidge CD, Baugh KE (1999) Mapping the artificial sky brightness in Europe from DMSP satellite measurements: the situation of the night sky in Italy in the last quarter of century. Astrophysics 16:41:51
Clarke JA (1983) Moonlight's influence on predator/pray interactions between short-eared owls (*Asio flammeus*) and deermice (*Peromyscus maniculatus*). Behav Ecol Sociobiol 13:205–209
Clayton DG, Bernardinelli L, Montomoli C (1993) Spatial correlation in ecological analysis. Int J Epidemiol 22:1193–1202
Coetzee CG (1970) The relative tail length of striped mice Rhabdomys pumilio (Sparrman 1784), in relation to climate. Zool Afr 5:1–6
Cohen M, Lippman M, Chabner B (1978) Role of pineal gland in aetiology and treatment of breast cancer. The Lancet 2(8094):814–816
Cohen P, Wax Y, Modan B (1983) Seasonality in the occurrence of breast cancer. Cancer Res 43:892–896
Cohen P (1983) Cancer and seasonal patterns. Am J Epidemiol 118(5):785–786
Cohen P (1970) Comparison of seasonal influence on pregnancy duration in man and domestic animals. Int J Biometeorol 14(4):365–369
Cohen RA, Albers HE (1991) Disruption of human circadian and cognitive regulation following a discrete hypothalamic lesion: a case study. Neurology 41:726–729
Conlon M, Lightfoot N, Kreiger N (2007) Rotating shift work and risk of prostate cancer. Epidemiology 18:182–183
Copeland KT, Checkoway H, McMichael AJ, Holbrook RH (1977) Bias due to misclassification in the estimation of relative risk. Am J Epidemiol 105:488–495
Cos S, Mediavilla MD, Fernández R, González-Lamuño D, Sánchez-Barceló EJ (2002) Does melatonin induce apoptosis in MCF-7 human breast cancer cells in vitro? J Pineal Res 32(2):90–6
Coughlin SS (1990) Recall bias in epidemiologic studies. J Clin Epidemiol 43(1990):87–91
Czeisler CA, Shanahan TL, Klerman EB, Martens H, Brotman DJ, Emens JS, Klein T, Rizzo JF (1995) Suppression of melatonin secretion in some blind patients by exposure to bright light. N Engl J Med 332:6–11
Davis S, Mirick DK, Chen C, Stanczyk FZ (2006) Effects of 60-Hz magnetic field exposure on nocturnal 6-sulfatoxymelatonin, estrogens, luteinizing hormone, and follicle-stimulating hormone in healthy reproductive-age women: results of a crossover trial. Ann Epidemiol 16:622–631

Davis S, Mirick DK, Stevens RG (2001) Night shift work, light at night, and risk of breast cancer. J Natl Cancer Inst 93:1557–1562

de Vries E, Boniol M, Severi G, Eggermont AM, Autier P, Bataille V, Doré JF, Coebergh JW (2005) Public awareness about risk factors could pose problems for case-control studies: the example of sunbed use and cutaneous melanoma. Eur J Cancer 41:2150–2154

DeCoursey PJ (1986) Light-sampling behavior in photoentrainment of a rodent circadian rhythm. J Comp Physiol 159A:161–169

Demers PA, Thomas DB, Rosenblatt KA, Jimenez LM, McTiernan A, Stalsberg H, Stemhagen A, Thompson WD, McCrea-Curnen MG, Satariano W, Austin DF, Isacson P, Greenberg RS, Key C, Kolonel LN, West DW (1991) Occupational exposure to electromagnetic fields and breast cancer in men. Am J Epidemiol 134(4):340–347

Di Laura D. et al (eds) The lighting handbook, 10th edn

Dijk DJ, Cajochen C, Borbély AA (1991) Effect of a single 3-hour exposure to bright light on core body temperature and sleep in humans. Neurosci Lett 121(1–2):59–62

Dijk DJ, Visscher CA, Bloem GM, Beersma DGM, Daan S (1987a) Reduction of human sleep duration after bright light exposure in the morning. Neurosci Lett 73:181–186

DMSP (2004) DMSP Nighttime lights data download

Doll CNH (2011) Population detection profiles of DMSP-OLS night-time imagery by regions of the world. Proceedings of the 30th Asia-Pacific advanced network meeting, pp 191–207

Elliott P, Wartenberg D (2004) Spatial epidemiology: current approaches and future challenges. Environ Health Perspect 112(9):998–1006

Elliott P. Cuzick J, English D, Stern R. (eds) (1992; 1996 reprint) Geographical and environmental epidemiology. Methods for small area studies, Oxford University press, New York, pp 404

Elvidge CD, Sutton PC, Tuttle BT, Ghosh T, Baugh KE (2009a) Global urban mapping based on nighttime lights. In: Gamba P, Herold M (eds) Global mapping of human settlement: experiences, datasets and prospects, CRC Press, Boca Raton, pp 129–145

Elvidge CD, Ziskin D, Baugh K E, Tuttle BT, Ghosh T, Pack DW, Erwin EH, Zhizhin M (2009c) A fifteen year record of global natural gas flaring derived from satellite data. Energies 2(3):595–622

Elvidge CD, Sutton PC, Ghosh T, Tuttle BT, Baugh KE, Bhaduri B, Bright E (2009b) A global poverty map derived from satellite data. Comput Geosci 35(8):1652–1660

ESRI (2007) ARCGIS, ESRI

Erren TC, Erren M, Lerchl A, Meyer-Rochow VB (2008) Clockwork blue: on the evolution of non-image-forming retinal photoreceptors in marine and terrestrial vertebrates. Naturwissenschaften 95:273–279

Falchi F, Cinzano P, Elvidge CD, Keith DM, Haim A (2011) Limiting the impact of light pollution on human health, environment and stellar visibility. J Environ Manage 92:2714–2722

Felsenstein D, Portnov BA (eds) (2005) Regional disparities in small countries. Springer, Heidelberg

Fransson E, De Faire U, Ahlbom A, Reuterwall C, Hallqvist J, Alfredsson L (2004) The risk of acute myocardial infarction: interactions of types of physical activity. Epidemiology 15:573–582

Friedenrich C, Howe G, Miller A (1991) The effect of recall bias on the association of calorie providing nutrients and breast cancer. Epidemiology 2:424–429

Gal G, Loew ER, Rudstam LG, Mohammadian AM (1999) Light and diel vertical migration: spectral sensitivity and light avoidance by *Mysis relicta*. Can J Fish Aquat Sci 56:311–322

García T, Esparza JL, Giralt M, Romeu M, Domingo JL, Gómez M (2010) Protective role of melatonin on oxidative stress status and RNA expression in cerebral cortex and cerebellum of AbetaPP transgenic mice after chronic exposure to aluminum. Biol Trace Elem Res 135:220–232

Gefeller O (2009) Invited commentary: recall bias in melanoma—much ado about almost nothing? Am J Epidemiol 169(3):267–270

Geonames (2008) Geonames world database. http://download.geonames.org/export/. Accessed 1 Feb 2008

Gerkema MP, Van der Zee EA, Feitsma LE (1994) Expression of circadian rhythmicity correlates with the number of argenin-vasopresine-immunoreactive cells in the suprachiasmatic nucleus of common vole, *Microtus arvalis*. Brain Res 639:93–101

Glaeser EL, Kallal HD, Scheinkman JA, Shleifer A (1992) Growth in cities. J Political Econ 100(6):1126–1152

Goodchild MF, Sun G, Yang S (1992) Development and test of an error model for categorical data. Int J Geogr Inf Syst 6(2):87–104

Gotway C, Young LJ (2002) Combining incompatible spatial data. J Am Stat Assoc 97(48):632–647

Gram-Hansenn K, Petersen NK (2004) Different everyday lives-different patterns of electrical use. ACEEE summer study on energy efficiency in buildings 2004, Pacific Grove, California, p 13

Greenland S (2001) Ecologic versus individual-level sources of bias in ecologic estimates of contextual health effects. Int J Epidemiol 30(6):1343–1350

Greenland S, Morgenstern H (1989) Ecological bias, confounding, and effect modification. Int J Epidemiol 18(1):269–274

Greenland S, Robins J (1994) Invited commentary: ecological studies—biases, misconceptions, and counterexamples. Am J Epidemiol 139(8):747–760

Griffith DA (2003) Spatial autocorrelation and spatial filtering: gaining understanding through theory and scientific visualization. Springer, Berlin, p 247

Guerrero JM, Reiter RJ (2002) Melatonin-immune system relationships. Curr. Top Med. Chem 2:167–179

Hahn RA (1991) Profound bilateral blindness and the incidence of breast cancer. Epidemiology 2:208–210

Haim A, Fourie le FR (1980a) Heat production in cold and long scotophase acclimated and winter acclimatized rodents. Int J Biometeorol 24:231–235

Haim A, Fourie le FR (1980b) Long scotophase increases heat production in Rhabdomys pumilio and Praomys natalensis (Rodentia). S Afr J Sci 76:89

Haim A, Fourie le FR (1982) Effects of melatonin on heat production and enzymatic activity in diurnal and in nocturnal rodents. Comp Biochem Phys 71A:473–475

Haim A, Zisapel, N (1995) Oxygen consumption and body temperature rhythms in the golden spiny mouse, response to changes in day length. Physiol Behav 58:775–778

Haim A, Yahav S (1982) Non-shivering thermogenesis in winter acclimatized and in long scotophase and cold acclimated *Apodemus mystacinus* (Rhodentia). J Therm Biol 7:193–195

Haim A, Brandes O, Afik D, Zisapel N (2001) Light manipulations as a possibility for outbreak-control of the vole microtus socialis. In: Pelz HJ, Cowan, DP, Feare CJ (eds) Advances in Vertebrate Pest Management II, Filander Verlag, Furth, pp 331–336

Haim A, Shanas U, Zisapel N, Gilboa A (2004) Rodent pest control: the use of photoperiod manipulations. Adv Vertebr Pest Manage 3:29–38

Haim A, Heth G, Pratt H, Nevo E (1983) Photoperiodic effects on thermoregulation in a 'blind' subterranean mammal. J Exper Biol 107:59–64

Haim A, Saarela S, Hissa R (1979) Heat production induced by photoperiodicity in the pigeon. Comp Biochem Physiol 63A:647–649

Haim A, Shanas U, Zubidad AS, Scantelbry M (2005) Seasonality and seasons out of time—the thermoregulatory effects of light interference. Chronobiol Int 22:57–64

Haim A, Shanas U, Zisapel N, Gilboa A (2004) Rodent pest control: the use of photoperiod manipulations. Adv Vertebrate Pest Manage III:29–38

Haim A, Youkler A, Harel O, Schwimmer H, Fares F (2010) Chronobiology affects prostate cancer cells growth *in vivo*. Sleep Sci 3:32–35

Haim A (1982) Effects of long scotophase and cold acclamation on heat production in two diurnal rodents. J Comp Physiol B148:77–81

Haim A, Portnov BA (2011) LAN and breast cancer risk: can we see a forest through the trees? In: Rea M,Brons J, Figueiro M (eds) Measurements of light at night (LAN) for a sample of women school teachers. Chronobiology International 28(8):734–736

Haim A, Kloog I, Rennert HS, Portnov BA (2011) Light pollution in the sleeping habitat: are eyelids impassible barriers to LAN? In: Schernhammer E, Stone K (eds) Light Pollution ≠ Light Pollution? Chronobiology International 28(4):379–380

Han X, Naeher LP (2006) A review of traffic-related air pollution exposure assessment studies in the developing world. Environ Int 32:106–120

Haney JF (1993) Environmental control of diel vertical migration behavior. Archiv fur Hydrobiologie. Biebefte. Ergebnisse der Limnologie 39:1–17

Hansen J (2001a) Light at night, shiftwork, and breast cancer risk. J Natl Cancer Inst 93:1513–1515

Hansen J (2001b) Increased breast cancer risk among women who work predominately at night. Epidemiology 12:74–77

Hardeland R, Pandi-Perumal SR, Cardinali DP (2006) Molecules in focus melatonin. Int J Biochem Cell Biol 38:313–316

Hardeland R, Poeggeler B (2003) Minireview Non-vertebrate melatonin. J Pineal Res 34:233–241

Hardeland R, Coto-Montes A, Poeggeler B (2003) Circadian rhythms, oxidative stress, and antioxidative defense mechanisms. Chronobiol Int 20:921–962

Hastings MH, Reddy AB, Maywood ES (2003) A clockwork web: circadian timing in brain and periphery, in health and disease. Nat Rev Neurosci 4:649–661

Hätönen T (2000) The impact of light on the secretion of melatonin in humans. Academic Dissertation (presented for public discussion at the Institute of Biomedicine, Department of Physiology, Helsinki, on April 28, 2000)

Hatori M, Hirota T, Litsuka M, Kurabayashi N, Haraguchi S, Kokame K, Sato R, Nakai A, Miyata T, Tsutsul K (2011) Light-dependent and circadian clock-regulated activation of sterol regulatory element binding protein, X-box-binding protein 1, and heat shock factors pathways. Proc Natl Acad Sci USA 108:4864–4869

Haus E (2007) Circadian disruption in shiftwork is probably carcinogenic to humans. Chronobiol Int 24:1255–1256

Heldmaier G, Steinlechner S, Rafael J, Visiansky P (1981) Photoperiodic control and effects of melatonin on non-shivering thermogenesis and brown adipose tissue. Science 212:917–919

Hertzog ED, Takahashi JS, Block GD (1998) Clock controls circadian period in isolated superachiasmatic nucleus neurons. Nat Neurosci 1:708–713

Heuvelink GBM, Burrough PA (1989) Propagation of errors in spatial modeling with GIS. Int J Geogr Inf Syst 3(4):303–322

Hill AB (1965a) The environment and disease: association or causation? Proc R Soc Med 58:295–300

Hoffmann K (1979) Photoperiodic effects in the Djungarian hamster: one minute of light during darktime mimics influence of long photoperiods on testicular recrudescence, body weight and pelage colour. Experientia 35:1529–1530

Hollan J (2004) Metabolism-influencing light: measurement by digital cameras, Poster at Cancer and Rhythm Oct 14–16, 2004. Graz, Austria

Hrushesky WJ, Teslow T, Halberg F, Kiang D, Kennedy BJ (1979) Temporal components of predictable variability along the 1-year scale in estrogen receptor concentration of primary human breast cancer. Proc Am Assoc Cancer Res 20:331

Hulshof KF, Brussaard JH, Kruizinga AG, Telman J, Lowik MR (2003) Socio-economic status, dietary intake and 10 y trends: the Dutch National Food Consumption Survey. Eur J Clin Nutr 57:128–137

Hulshof KF, Lowik MR, Kok FJ, Wedel M, Brants HA, Hermus RJ, ten Hoor F (1991) Diet and other life-style factors in high and low socio-economic groups (Dutch Nutrition Surveillance System). Eur J Clin Nutr 45:441–450

Infante-Rivard C, Jacques L (2000) An empirical study of parental recall bias. Am J Epidemiol 152(5):480–486
Jansky L (1973) Nonshivering thermogenesis and its thermoregulatory significance. Biol Rev 48:85–132
Jean-Louis G, Kripke DF, Cole RJ, Elliott JA (2000) No melatonin suppression by illumination of popliteal fossae or eyelids. J Biol Rhythms 15:265–269
Jumbe BLC (2004) Cointegration and causality between electricity consumption and GDP: empirical evidence from Malawi. Energy Econ 26:61–68
Kakizaki M, Inoue K, Kuriyama S, Sone T, Matsuda-Ohmori K, Nakaya N, Fukudo S, Tsuji I (2008) Sleep duration and the risk of prostate cancer: the Ohsaki Cohort study. Br J Cancer 99:176–178
Kelsey JL, Gammon MD (1990) The epidemiology of breast cancer. Cancer J Clin 41:146–165
Khalsa SBS, Jewett ML, Cajochen C, Czeisler CA (2003) A phase response curve to single bright light pulses in human subjects. J Physiol 549:945–952
Kinnear PR, Gray CD (2007) SPSS 15 made simple. Psychology Press, Hove, p 620
Kiran Chand TR, Badarinath KVC, Krishna Prasad V, Murthy MSR, Elvidge CD, Tuttle BT (2006) Monitoring forest fires over the Indian region using defense meteorological satellite program-operational linescan system nighttime satellite data. Remote Sens Environ 103(2):165–178
Kliukiene J, Tynes T, Andersen A (2003) Follow-up of radio and telegraph operators with exposure to electromagnetic fields and risk of breast cancer. Eur J Cancer Prev 12(4):301–307
Kloog I, Haim A, Stevens RG, Barchana M, Portnov BA (2008) Light at night co-distributes with incident breast but not lung cancer in the female population of Israel. Chronobiol Int 25:65–81
Kloog I, Haim A, Portnov BA (2009a) Using kernel density function as an urban analysis tool: Investigating the association between nightlight exposure and the incidence of breast cancer in Haifa, Israel. Comput Environ Urban Sys 33:55–63
Kloog I, Haim A, Stevens RG, Portnov BA (2009b) Global co-distribution of light at night (LAN) and cancers of prostate, colon, and lung in men. Chronobiol Int 26:108–125
Kloog I, Stevens R, Haim A, Portnov B (2010) Nighttime light level co-distributes with breast cancer incidence worldwide. Cancer Causes Control 1–10
Kloog I, Portnov BA, Rennert HS, Haim A (2009c) Does the modern urbanized sleeping habitat pose a breast cancer risk? Chronobiol Int 28(1):76–80
Kohidai L, Vakkuri O, Keresztesi M, Leppaluoto J, Casaba G (2002) Melatonin in the unicellular Tetrahymena pyriformis: effects of different lighting conditions. Cell Biochem Func 20:269–272
Kolar J, Machackova I, Eder J, Prinsen E, Dongen WV, Onckelen HV Illnerova H (1997) Melatonin: occurrence and daily rhythm in Chenopodium rubrum. Phytochemistry 44:1407–1413
Konopka RJ, Benzer S (1971) Clock mutants of *Drosophila melanogaster*. Proc Natl Acad Sci USA 68:2112–2116
Koolhaas JM, Bartolomucci A, Buwalda B, Boer SFD, Flügge G, Korte SM, Meerlo P, Murison R, Olivier B, Palanza P et al (2011) Stress revisited: a critical evaluation of the stress concept. Neurosci Biobehav R 35:1291–1301
Korkmaz A, Reiter RJ (2008) Epigenetic regulation: a new research area for melatonin? J Pineal Res 44:41–44
Kregel KC, Zhang HJ (2007) An integrated view of oxidative stress in aging: basic mechanisms, functional effects, and pathological considerations. Am J Physiol Regul Integr Comp Physiol 292:R18–R36
Kregel KC (2002) Heat shock proteins: modifying factors in physiological stress responses and acquired thermotolerance. J Appl Physiol 92:2177–2186

Krieger N, Chen JT, Waterman PD, Soobader MJ, Subramanian SV, Carson R (2002) Geocoding and monitoring of US socioeconomic inequalities in mortality and cancer incidence: does the choice of area-based measure and geographic level matter? The public health disparities geocoding project. Am J Epidemiol 156:471–482

Kubo T, Ozasa K, Mikami K, Wakai K, Fujino Y, Watanabe Y, Miki T, Nakao M, Hayashi K, Suzuki K, Mori M, Washio M, Sakauchi F, Ito Y, Yoshimura T, Tamakoshi A (2006a) Prospective cohort study of the risk of prostate cancer among rotating-shift workers: findings from the Japan collaborative cohort study. Am J Epidemiol 164:549–555

Lerner AR, Case JD, Takahashi Y (1958) Isolation of melatonin, a pineal factor lightens melanocytes. J Am Chem Soc 80:2057–2058

Lewy AJ, Wehr TA, Goodwin FK, Newsome DA, Markey SP (1980) Light suppresses melatonin secretion in humans. Science 210:1267–1269

Longcore T, Rich C (2004) Ecological light pollution. Front Ecol Environ 2:191–198

Lynch GR (1970) Effect of photoperiod and cold acclimation on non-shivering thermogenesis in *Peromyscus leucopus*. Am Zool 10:308

Madison T, Schottenfeld D, James SA, Schwartz AG, Gruber SB (2004) Endometrial cancer: socioeconomic status and racial/ethnic differences in stage at diagnosis, treatment, and survival. Am J Public Health 94:2104–2111

Maheswaran R, Strachan DP, Dodgeon B, Best NG (2002) A population-based case-control study for examining early life influences on geographical variation in adult mortality in England and Wales using stomach cancer and stroke as examples. Int J Epidemiol 31:375–382

Mahoney M, Bult A, Smale L (2001) Phase response curve and light-induced fos expression in the suprachiasmatic nucleus and adjacent hypothalamus of *Arvicanthis niloticus*. J Biol Rhythms 16:149–162

Markus RP, Ferreira ZS, Fernandes PACM, Cecon E (2007) The immune-pineal axis: a shuttle between endocrine and paracrine melatonin sources. Neuroimmunomodulation 14(3–4):126

Markus RP, Ferreira ZS (2011) The immune-pineal axis: the role of pineal and extra-pineal melatonin in modulating inflammation. Adv Neuroimmune Biol 1:95–104

Matés JM, Pérez-Gómez C, Núñez de Castro I (1999) Antioxidant enzymes and human diseases. Clin Biochem 32:595–603

Mawson AR (1998) Breast cancer in female flight attendants (letter). Lancet 352:626

McCord JM, Fridovich I (1969) Superoxide dismutase. An enzymic function for erythrocuprein (hemocuprein). J Biol Chem 244:6049–6055

McCoy J, Johnston K (2001) Using ArcGIS spatial analyst. ESRI, Redlands

McIntyre IM, Norman TR, Burrows GD, Armstrong SM (1989a) Human melatonin suppression by light is intensity dependent. J Pineal Res 6:149–156

McNerney J, Farmer JD, Trancik JE (2011) Historical costs of coal-fired electricity and implications for the future. Energy Policy 39(6):3042–3054

Miller BA, Chu KC, Hankey BF, Ries LAG (2008) Cancer incidence and mortality patterns among specific Asian and Pacific Islander populations in the US. Cancer Causes Control 19:227–256

Minami M, ESRI (2000) Using ArcMap: GIS. ESRI, Redlands, pp 54

Minors DS, Waterhouse JM (1981) Anchor sleep as a synchronizer of rhythms on abnormal routines. Int J Chronobiol 7:165–188

Mohr PJ, Taylor BN, and DB Newell (2011) The 2010 CODATA recommended values of the fundamental physical constants (Web Version 6.3). Database developed by J. Baker, M. Douma, and S. Kotochigova. Available. http://physics.nist.gov/constants (Accessed 1 Aug 2012). National Institute of Standards and Technology, Gaithersburg, MD 20899

Moldofsky H, Lue FA, Davidson JR et al (1989) Effects of sleep deprivation on human immune function. FASEB J 3:1972–1977

Monteleone P, Maj M, Fuschino A, Kemali D (1992) Physical stress in the middle of the dark phase does not affect light-depressed plasma melatonin levels in humans. Neuroendocrinology 55:367–371

Moore CB, Siopes TD (2000) Effects of lighting conditions and melatonin supplementation on the cellular and hormonal immune response in Japanese quail Coturnix coturnix japonica. Gen Comp Endocrinol 119:95–104

Moore MV, Pierce SM, Walsh HM, Kvalvik KK, Lim JD (2001) Urban light pollution alters the diel vertical migration of Daphnia. Int Verein Limnol 27:1–4

Moore RM, Eichler VB (1972) Loss of circadian adrenal corticosterone rhythm following suprachiasmatic lesions in the rat. Brain Res 42:201–206

Moore-Ede MC, Sulzman FM, Fuller CA (1982) The clocks that time us. Harvard University Press, Cambridge

Morgenstern H, Thomas D (1993) Principles of study design in environmental epidemiology. Environ Health Perspect 101(Suppl 4):23–38

Navara KJ, Nelson RJ (2007) The dark side of light at night: physiological, epidemiological and ecological consequences. J Pineal Res 43:215–224

Nelson RJ, Drazen DL (1999) Melatonin mediates seasonal adjustments in immune function. Reprod Nutr Dev 39:383–398

Nelson RJ (2004) Seasonal immune function and sickness responses. Trends Immunol 25:187–192

Nelson RJ (2005) An introduction to behavioral endocrinology Sinauer associates, Inc, Sunderland, Massachusetts

NHMFL (2011) Electricity timeline. (http://www.magnet.fsu.edu/education/. Accessed Aug 2012)

NYT (2011) Inventors. (http://inventors.about.com/library/. Accessed Aug 2012)

O'Leary ES, Vena JE, Freudenheim JL, Brasure J (2004) Pesticide exposure and risk of breast cancer: a nested case-control study of residentially stable women living on Long Island. Environ Res 94:134–144

O'Leary ES, Schoenfeld ER, Stevens RG, Kabat GC, Henderson K, Grimson R, Gammon MD, Leske MC (2006) Shift work, light at night, and breast cancer on Long Island, New York. Am J Epidemiol 164:358–366

Openshaw S (1984) The modifiable areal unit problem. In: Concepts and techniques in modern geography. Monograph Series. Geo Books, London, No. 38, p 41

Parkin DM, Bray F, Ferlay J, Pisani P (2001a) Estimating the world cancer burden: Globocan 2000. Int J Cancer 94:153–156

Parkin DM, Bray F, Ferlay J, Pisani P (2005) Global cancer statistics, 2002. CA Cancer J Clin 55:74–108

Parkin DM, Bray FI, Devesa SS (2001b) Cancer burden in the year 2000. The global picture. Eur J Cancer 37(Suppl 8):S4–S66

Parkin DM, Wabinga H, Nambooze S (2001c) Completeness in an African cancer registry. Cancer Causes Control 12:147–152

Parr CL, Hjartåker A, Laake P, Lund E, Veierød MB (2009) Recall bias in melanoma risk factors and measurement error effects: a nested case-control study within the Norwegian women and cancer study. Am J Epidemiol 169:257–266

Pauley SM (2004) Lighting for the human circadian clock: recent research indicates that lighting has become a public health issue. Med Hypotheses 63:588–596

Pekkanen J, Pearce N (2001) Environmental epidemiology: challenges and opportunities. Environ Health Perspect 109:1–5

Piggins HD, Antle MC, Rusak B (1995) Neuropeptides phase shift the mammalian circadian pacemaker. J Neurosci 15:5612–5622

Poeggeler B, Thuermann S, Dose A, Schoenke M, Burkhardt S, Hardeland R (2002) Melatonin's unique radical scavenging properties—roles of its functional substituents as revealed by a comparison with its structural analogs. J Pineal Res 33:20–30

Portnov BA, Dubnov J, Barchana M (2007) On ecological fallacy, assessment errors stemming from misguided variable selection, and the effect of aggregation on the outcome of epidemiological study. J Expo Sci Environ Epidemiol 17:106–121

Pukkala E, Auvinen A, Wahlberg G (1995) Incidence of cancer among Finnish airline cabin attendants, 1967–92. Br Med J 311:649–652

Ralph MR, Menaker M (1988) A mutation of the circadian system in golden hamster. Science 241:1125–1127

Raphael K (1987) Recall bias: a proposal for assessment and control. Int J Epidemiol 16(2): 167–170

Rea MS (ed) (2000) The IESNA Lighting Handbook, 9th edn, Illuminating Engineering Society of North America, ISBN: 978-0879951504

Rea MS, Brons JA, Figueiro MG (2011) Measurements of light at night (LAN) for a sample of female school teachers. Chronobiol Int 28(8):673–680

Rechencq M, Sosnovsky A, Macchi PJ, Alvear PA, Vigliano PH (2011) Extensive diel fish migrations in a deep ultraoligotrophic lake of Patagonia Argentina. Hydrobiologia 658(1):147–161

Refii-El-Idrissi M, Calvo JR, Giordano M et al (1996) Specific binding of 2-[125I]iodomelatonin by rat spleen crude membranes: day-night variations and effect of pinealectomy and continuous light exposure. J Pineal Res 20:33–38

Reiter RJ, Fraschini F (1969) Endocrine aspects of the mammalian pineal gland: a review. Neuroendocrinology 5:219–255

Reiter RJ, Tan DX, Manchester LC, Terron MP, Flores LJ, Koppisetti S (2007) Medical implications of melatonin: receptor mediated and receptor independent actions. Adv Med Res 52:11–28

Reiter RJ, Tan DX, Mayo JC, Sainz RM, Leon J, Czarnocki Z (2003) Melatonin as an antioxidant: biochemical mechanisms and pathophysiological implications in humans. Acta Biochim Pol 50:1129–1146

Reiter RJ (1993) The melatonin rhythm: both a clock and a calendar. Experientia 49:654–664

Reppert SM, Weaver DR (2002) Coordination of circadian timing in mammals. Nature 418:935–941

Rezaeian M, Dunn G, St Leger S, Appleby L (2006) Ecological association between suicide rates and indices of deprivation in the north west region of England: the importance of the size of the administrative unit. J Epidemiol Community Health 60:956–961

Rich C, Longcore T (2005) Ecological consequences of artificial night lighting. Island Press, Washington

Robinson WS (1958) Ecological correlations and the behavior of individuals. Am Sociol Rev 15:351–357

Roenneberg T, Lucas RJ (2002) Light, endocrine systems, and cancer—a view from circadian biologists. Neuroendocrine Lett 23(suppl 2):82–83

Ronenberg T, Merrow M (2001) Circadian systems: different levels of complexity. Philos Trans R Soc Lond B Biol Sci 356(1415):1687–1696

Roosli M, Lortscher M, Egger M, Pfluger D, Schreier N, Lortscher E, Locher P, Spoerri A, Minder C (2007) Leukaemia, brain tumours and exposure to extremely low frequency magnetic fields: cohort study of Swiss railway employees. Occup Environ Med 64:553–559

Rothman K, Greenland S, Lash T (2008) Modern epidemiology. Lippincott Williams and Wilkins, Philadelphia

Rydell J (2005) Bats and their insect prey at streetlights. In: Ecological consequences of artificial night lighting (Rich & Longcore Eds.). Islandpress, Washington, pp 43–60

Sasseville A, Paquet N, sevigny J, Herbert M (2006) Blue blocker glasses impede the capacity of bright light to suppress melatonin production. J Pineal Res

Schernhammer ES, Laden F, Speizer FE, Willett WC, Hunter DJ, Kawachi I, Colditz GA (2001) Rotating night shifts and risk of breast cancer in women participating in the nurses' health study. J Natl Cancer Inst 93:1563–1568

Schernhammer ES, Stone KL (2011) Light pollution ≠ Light pollution? Chronobiol Int 28(4):378–379

Schwartz WJ, Bosis NA, Hedley-Whyte ET (1986) A discrete lesion of ventral hypothalamus and optic chiasm that distributed the daily temperature rhythm. J Neurol 233:1–4

Schwartzet WJ, Gainer H (1997) Suprachiasmatic nucleus: use of ^{14}C-labeled deoxyglucose uptake as a functional marker. Science 197:1089–1091

Scott D, Curtis B, Twumasi FO (2002) Towards the creation of a health information system for cancer in KwaZulu-Natal, South Africa. Health Place 8:237–249

Selvin HC (1958) Durkheim's 'suicide' and problems of empirical research. Am J Sociol 63:607–619

Selye H (1950) Stress and the general adaptation syndrome. BMJ 1:1383–1392

Sharpe LT, Stockman A, Jagla W, Jägle H (2005) A luminous efficiency function, V*(lambda), for daylight adaptation. J Vis 21, 5(11):948–968

Sherman LP, Zylka MJ, Weaver DR, Kolakowski LF, Reppert SM (1997) Two periods homologs: circadian expression and photic regulation in the suprachiasmatic nuclei. Neuron 19:1261–1269

Small C, Elvidge CD, Balk D, Montgomery M (2011) Spatial scaling of stable night lights. Remote Sens Environ 115:269–280

Stephan FK, Zuker I (1972) Circadian rhythms in drinking behavior and locomotor activities are eliminated by hypothalamic lesions. Proc Natl Acad Sci USA 69:1583–1586

Stevens RG, Rea MS (2001a) Light in the built environment: potential role of circadian disruption in endocrine disruption and breast cancer. Cancer Causes Control 12:279–287

Stevens RG (2005a) Circadian disruption and breast cancer: from melatonin to clock genes. Epidemiology 16:254–258

Stevens RG (2011) Testing the light-at-night (LAN) theory for breast cancer causation. Chronobiol Int 28(8):653–656

Stevens RG, Davis S, Thomas DB, Anderson LE, Wilson BW (1992) Electric power, pineal function, and the risk of breast cancer. FASEB J 6:853–860

Stevens RG, Davis S (1996) The melatonin hypothesis: electric power and breast cancer. Environ Health Perspect 104(Suppl 1):135–140

Stevens RG, Rea MS (2001b) Light in the built environment: potential role of circadian disruption in endocrine disruption and breast cancer. Cancer Causes Control 12:279–287

Stevens RG, Blask DE, Brainard GC, Hansen J, Lockley SW, Provencio I, Rea MS, Reinlib L (2007) Meeting report: the role of environmental lighting and circadian disruption in cancer and other diseases. Environ Health Perspect 115:1357–1362

Stott DH (1958) Some psychosomatic aspects of casualty in reproduction. J Psychosom Res 3(1):42–55

Straif K, Baan R, Grosse Y, Secretan B, El Ghissassi F, Bouvard V, Altieri A, Benbrahim-Tallaa L, Cogliano V (2007) Carcinogenicity of shift-work, painting, and fire-fighting. Lancet Oncol 8:1065–1066

Suvanto S, Harma M, Ilmarinen J, Partinen M (1993) Effects of 10 h time zone changes on female flight attendants' circadian rhythms of body temperature, alertness and visual search. Ergonomics 36:613–625

Tal O, Haim A, Harel O, Gerchman Y (2011) Melatonin as an antioxidant and its semi-lunar rhythm in green macroalga Ulva Sp. J Exp Bot 62:1903–1910

Tamarkin L, Danforth D, Lichter A, DeMoss E, Cohen M, Chabner B, Lippman M (1982) Decreased plasma melatonin peak in patients with estrogen receptor positive breast cancer. Science 216:1003–1005

Tan DX, Reiter RJ, Manchester LC, Yan MT, El-Sawi M, Sains RM, Mayo JC, Kohen R, Allegra M, Hardeland R (2002) Chemical and physical properties and potential mechanisms: melatonin is a broad spectrum antioxidant and free radical scavenger. Curr Top Med Chem 2:181–197

Tynes T, Hannevik M, Andersen A, Vistnes A, Haldorsen T (1996) Incidence of breast cancer in Norwegian female radio and telegraph operators. Cancer Causes Control 7:197–204

Van den Heiligenberg S, Depres-Brummer P, Barbason H, Claustrat B, Reynes M, Levi F (1999) The tumor promoting effect of constant light exposure on diethylnitrosamine-induced hepatocarcinogenesis in rats. Life Sci 64:2523–2534

Vaughan MK, Hubbard GB, Champney TH et al (1987) Splenic hypertrophy and extramedullary ematopoiesis induced in male Syrian hamsters by short photoperiod or melatonin injections and reversed by melatonin pellets or pinealectomy. Am J Anat 179:131–136

Verkasalo PK, Pukkala E, Stevens RG, Ojamo M, Rudanko SL (1999) Inverse association between breast cancer incidence and degree of visual impairment in Finland. Br J Cancer 80:1459–1460

Volzke H, Neuhauser H, Moebus S, Baumert J, Berger K, Stang A, Ellert U, Werner A, Doring A (2006) Urban-rural disparities in smoking behaviour in Germany. BMC Public Health 6:146

Vos JJ (1978) Colorimetric and photometric properties of a 2-deg fundamental observer. Color Res Appl 3:125–128

Vrijheid M, Armstrong BK, Bedard D, Brown J, Deltour I, Iavarone I, Krewski D, Lagorio S, Moore S, Richardson L, Giles GG, McBride M, Parent ME, Siemiatycki J, Cardis E (2009) Recall bias in the assessment of exposure to mobile phones, J Expo Sci Environ Epidemiol 19(4):369−381

Wagner S, Castel M, Gainer H, Yarom Y (1997) GABA in the mammalian suprachiasmatic nucleus and its rule in diurnal rhythmicity. Nature 387:598–603

Wakefield J, Shaddick G (2005) Health-exposure modeling and the ecological fallacy. Biostatistics. 7(3):438–445

Wehr TA (1992a) In short photoperiods, human sleep is biphasic. J Sleep Res 1:103–107

Wells BL, Horm JW (1992) Stage at diagnosis in breast cancer: race and socioeconomic factors. Am J Public Health 82:1383–1385

Welsh DK, Logothetis DE, Meister M, Reppert SM (1995) Individual neurons dissociated from suprachismatic nucleus express independently phase circadian firing rhythms. Neuron 14:697–706

Wever RA (1989) Light effects on human circadian rhythms: A review of recent Andechs experiments. J Biol Rhythms 4:161–185

Wright HR, Lack LC, Kennaway DJ (2004) Differential effects of light wavelength in phase advancing the melatonin rhythm. J Pineal Res 36:140–144

Young MW (2000) Marketing time for a kingdom. Science 288:451–453

Zeitzer JM, Dijk DJ, Kronauer R, Brown E, Czeisler C (2000) Sensitivity of the human circadian pacemaker to nocturnal light: melatonin phase resetting and suppression. J Physiol 526:695–702

Zisapel N, Tarrasch R, Laudon M (2005) The relationship between melatonin and cortisol rhythms: clinical implications of melatonin therapy. Drug Dev Res 65:119–125

Zubidat A, Ben-Shlomo R, Haim A (2007) Thermoregulatory and endocrine responses to light pulses in short-day acclimated social voles (Microtus socialis). Chronobiol Int 24:269–288

Zubidat A, Nelson RJ, Haim A (2009) Photosensitivity to different light intensities in blind and sighted rodents. J Exper Biol 212:3857–3864

Zubidat A, Nelson RJ, Haim A (2011) Spectral and duration sensitivity to light-at-night in 'blind' and sighted rodent species. J Exp Biol 214:3206–3217

Additional Reading

Anisimov VN (2005) Cancer in rodents: does it tell us about cancer in humans? Nat Rev Cancer 5:807–819

Baldwin WS, Barrett JC (1998) Melatonin: receptormediated events that may affect breast and other steroid hormone-dependent cancers. Mol Carcinog 21:149–155

Balsalobre A, Damiola F, Schibler U (1998) A serum shock induces circadian gene expression in mammalian tissue culture cells. Cell 93:929–937

Bartsch C, Bartsch H, Buchberger A, Stieglitz A, Effenberger-Klein A, Kruse-Jarres JD et al (1999) Serial transplants of DMBA-induced mammary tumors in Fischer rats as a model system for human breast cancer. Oncology 56:169–176

Bartsch C, Bartsch H, Jain AK, Laumas KR, Wetterberg L (1981) Urinary melatonin levels in human breast cancer. J Neural Transm 52:281–294

Bartsch C, Bartsch H, Karenovics A, Franz H, Peiker G, Mecke D (1997) Nocturnal urinary 6-sulphatoxymelatonin excretion is decreased in primary breast cancer patients compared to age-matched controls and shows negative correlation with tumor-size. J Pineal Res 23:53–58

Benshoff HM, Brainard GC, Rollag MD et al (1987) Suppression of pineal melatonin in Peromyscus leucopus by different monochromatic wavelengths of visible and near-ultraviolet light (UV-A). Brain Res 420:397–402

Berson DM, Dunn FA, Takao M (2002) Phototransduction by retinal ganglion cells that set the circadian clock. Science 295:1070–1073

Blask DE, Dauchy RT, Sauer LA (2005b) Putting cancer to sleep at night: the neuroendocrine/circadian melatonin signal. Endocrine 27:179–188

Blask DE, Sauer LA, Dauchy RT (2002) Melatonin as a chronobiotic/anticancer agent: cellular, biochemical and molecular mechanisms of action and their implications for circadian-based cancer therapy. Curr Top Med Chem 2:113–132

Blot WJ, Lanier A, Fraumeni JF, Bender TR (1975) Cancer mortality among Alaskan natives. J Natl Cancer Inst 55:547–554

Brainard G, Lewy A, Menaker M et al (1985) Effect of light wavelength on the suppression of nocturnal plasma melatonin in normal volunteers. Ann NY Acad Sci 453:376–378

Brainard GC, Hanifin JP, Rollag MD, Greeson JM, Byrne B, Glickman G, Gerner E, Sanford B (2001b) Human melatonin regulation is not mediated by the three cone photopic visual system. J Clin Endocrinol Metab 86:433–436

Brainard GC, Hanifin JP (2005) Photons, clocks, and consciousness. J Biol Rhythms 20:314–325

Brainard GC, Lewy AJ, Menaker M, Frederickson RH et al (1988) Dose-response relationship between light irradiance and the suppression of melatonin in human volunteers. Brain Res 454:213–218

Brainard GC, Richardson BA, King TS et al (1984) The influence of different spectra on the suppression of pineal melatonin content in the Syrian Hamster. Brain Res 294:333–339

A. Haim and B. A. Portnov, *Light Pollution as a New Risk Factor for Human Breast and Prostate Cancers*, DOI: 10.1007/978-94-007-6220-6,

Brindley GS, Du Croz JJ, Rushton WA (1966) The flicker fusion frequency of the blue-sensitive mechanism of color vision. J Physiol 183:497–500
Bullough JD, Bierman A, Figueiro MG, Rea MS (2008) On melatonin suppression from polychromatic and narrowband light. Chronobiol Int 25:653–655
Bullough JD, Figueiro MG, Possidente BP, Parsons RH, Rea MS (2005) Additivity in murine circadian phototransduction. Zoolog Sci 22:223–227
Bullough JD, Rea MS, Figueiro MG (2006) Of mice and women: light as a circadian stimulus in breast cancer research. Cancer Causes Control 17:375–383
Cahill GM, Grace MS, Besharse JC (1991) Rhythmic regulation of retinal melatonin: metabolic pathways, neurochemical mechanisms, and the ocular circadian clock. Cell Mol Neurobiol 11:529–560
Cajochen C, Munch M, Kobialka S, Krauchi K, Steiner R, Oelhafen P et al (2005b) High sensitivity of human melatonin, alertness, thermoregulation, and heart rate to short wavelength light. J Clin Endocrinol Metab 90:1311–1316
Canaple L, Kakizawa T, Laudet V (2003) The days and nights of cancer cells. Cancer Res 63:7545–7552
Caplan LS, Schoenfeld ER, O'Leary ES, Leske MC (2000) Breast cancer and electromagnetic fields—a review. Ann Epidemiol 10:31–44
Cardinali DP, Larin F, Wurtman RJ (1972) Control of the rat pineal gland by light spectra. Proc Natl Acad Sci USA 69:2003–2005
Carrillo-Vico A, Garcia-Maurino S (2003) Melatonin counteracts the inhibitory effect of PGE-2 on IL-2 production in human lymphocytes via its mt1 membrane receptor. FASEB J 17:755–757
Carrillo-Vico A, Guerrero JM, Lardone P, Reiter RJ (2005) A review of the multiple ac-tions of melatonin on the immune system. Endocrine 27:189–200
Chada S, Ramesh R, Mhashilkear AM (2003) Cytokine and chemokine-based gene ther-apy for cancer. Curr Opin Mol Ther 5:463–474
Chung ST, Pease PL (1999) Effect of yellow filters on pupil size. Optom Vis Sci 76:59–62
Coleman MP, Reiter RJ (1992) Breast cancer, blindness and melatonin. Eur J Cancer 28:501–503
Cos S, Fernandez R, Guezmes A (1998) Sanchez- Barcelo EJ. Influence of melatonin on invasive and metastatic properties of MCF-7 human breast cancer cells. Cancer Res 58:4383–4390
Cos S, Martinez-Campa C, Mediavilla MD, Sanchez-Barcelo C (2005) Melatonin modu-lates aromatase activity in MCF-7 human breast cancer cells. J Pineal Res. 38:217–222
Cos S, Mediavilla D, Martinez-Campa C, Gonzalez A, Alonzo-Gonzalez C, San-chez-Barcelo EJ (2006) Exposure to light-at-night increases the growth of DMBA-induced mammary adenocarcinoma in rats. Cancer Lett 235:266–270
Czeisler CA, Allan JS, Strogatz SH et al (1986) Bright light resets the human circadian pacemaker independent of the timing of the sleep-wake cycle. Science 233:667–671
Daan S, Albrecht U, Van Der Horst GTJ et al (2001) Assembling a clock for all seasons: are there M and E oscillators in the genes. J Biol Rhythms 16:105–116
Dacey DM, Liao HW, Peterson BB, Robinson FR, Smith VC, Pokorny J, Yau KW, Gamlin PD (2005) Melanopsin-expressing ganglion cells in primate retina signal color and irradiance and project to the LGN. Nature 433:749–754
Dai J, Inscho EW, Yuan L, Hill SM (2002) Modulation of intracellular calcium and calmodulin by melatonin in MCF-7 human breast cancer cells. J Pineal Res 32:112–119
Damianaki A, Bakogeorgou E, Kampa M, Notas G, Hatzoglou A, Panagiotou S, Gemetzi C, Kouroumalis E, Martin PM, Castanas E (2000) Potent inhibitory action of red wine polyphenols on human breast cancer cells. J Cell Biochem 78:429–441
Danforth DN, Tamarkin L, Mulvihilli JJ, Bagley CS, Lippman ME (1985) Plasma melatonin and the hormone-dependency of human breast cancer. J Clin Oncol 3:941–948
Dauchy RT, Sauer LA, Blask ED, Vaughan GM (1997) Light contamination during the dark phase in "photoperiodically controlled" animal rooms: effect on tumor growth and metabolism in rats. Lab Anim Sci 47:511–518

Davidson AJ, Sellix MT, Daniel J, Yamazaki S, Menaker M, Block GD (2006) Chronic jet-lag increases mortality in aged mice. Curr Biol 16:R914–R916
Del Rio B, Garcia-Pedrero JM, Martinez-Campa C, Zuazua P, Lazo PS, Ramos S (2004) Melatonin, an endogenous-specific inhibitor of estrogen receptor alpha via calmodulin. J Biol Chem. 279:38294–38302
Dijk DJ, Cajochen C, Borbely AA (1991) Effect of a single 3-hour exposure to bright light on core body temperature and sleep in humans. Neurosci Lett 121:59–62
Dijk DJ, Visscher CA, Bloem GM, Beersma DGM, Daan S (1987) Reduction of human sleep duration after bright light exposure in the morning. Neurosci Lett 73:181–186
Dkhissi-Benyahya O, Gronfier C, De Vanssay W, Flamant F, Cooper HM (2007) Modeling the role of mid-wavelength cones in circadian responses to light. Neuron 53:677–687
Dunlap JC, Loros JJ, DeCoursey PJ (2004) Chronobiology—biological timekeeping. Sinauer Assoc. Inc., Sunderland
Eck KM, Duffy L, Ram PT, Ayettey S, Chen I, Cohn CS et al (1998) A sequential treatment regimen with melatonin and all-trans retinoic acid induces apoptosis in MCF-7 tumour cells. Br J Cancer 77:2129–2137
Erren TC, Meyer-Rochow VB, Erren M (2007) Health clues from polar regions. Science 316:540
Erren TC, Piekarski C (eds) (2002) Light, endocrine systems andcancer. Neuroendocrinol Lett 23(Suppl 2):1–104
Erren TC, Piekarski C (1999) Does winter darkness in the Artic protect against cancer? The melatonin hypothesis revisited. Med Hypotheses 53:1–5
Erren TC, Reiter RJ, Piekarski C (2003) Light, timing of biological rhythms, and chronodisruption in man. Naturwissenschaften 90:485–494
Escames G, Lopez LC, Ortiz F, Lopez A, Garcia JA, Ros E, Acuna-Castroviejo D (2007) Attenuation of cardiac mitochondria dysfunction by melatonin in septic mice. FEBS J 274:2135–2147
Feychting M, Osterlund B, Ahlbom A (1998) Reduced cancer incidence among the blind. Epidemiol 9:490–494
Figueiro MG, Bierman A, Bullough JD, Rea MS (2009a) A personal light-treatment device for improving sleep quality in the elderly: dynamics of nocturnal melatonin suppression at two exposure levels. Chronobiol Int 26:726–739
Figueiro MG, Bierman A, Plitnick B, Rea MS (2009b) Preliminary evidence that bothblue and red light can induce alertness at night. BMC Neurosci 10:105
Figueiro MG, Bierman A, Rea MS (2008) Retinal mechanisms determine the subadditive response to polychromatic light by the human circadian system. Neurosci Lett 438:242–245
Figueiro MG, Bullough JD, Bierman A, Fay CR, Rea MS (2007) On light as an alerting stimulus at night. Acta Neurobiol Exp 67:171–178
Figueiro MG, Bullough JD, Bierman A, Rea MS (2005) Demonstration of additivity failure in human circadian phototransduction. Neuroendocrinol Lett 26:493–498
Figueiro MG, Bullough JD, Parsons RH, Rea MS (2004) Preliminary evidence for spectral opponency in the suppression of melatonin by light in humans. Neuroreport 15:313–316
Figueiro MG, Rea MS, Bullough JD (2006a) Does architectural lighting contribute to breast cancer? J Carcinogenesis 5:20
Figueiro MG, Rea MS, Bullough JD (2006b) Circadian effectiveness of two polychromatic lights in suppressing human nocturnal melatonin. Neurosci Lett 406:293–297
Figueiro MG, Rea MS, Bullough JD (2006c) Does architectural lighting contribute to breast cancer? J Carcinog 5:20
Filipski E, Delaunay F, King VM, Wu MW, Claustrat B, Gréchez- Cassiau A et al (2004) Effects of chronic jet lag on tumor progression in mice. Cancer Res 64:7879–7885
Filipski E, Innominato PF, Wu MW, Li XM, Iacobelli S, Xian LJ et al (2005) Effects of light and food schedules on liver and tumor molecular clocks in mice. J Natl Cancer Inst 97:507–517
Filipski E, King VM (2002) Host circadian clock as a control point in tumor progression. J Natl Cancer Inst 94:690–697

Filipski E, Li XM (2006) Disruption of circadian coordination and malignant growth. Cancer Causes Control 17:509–514

Fu L, Lee CC (2003) The circadian clock: pacemaker and tumour suppressor. Nat Rev Cancer 3:350–361

Fu L, Pelicano H, Liu J, Huang P, Lee CC (2002) The circadian gene Period2 plays an important role in tumor suppression and DNA damage response *in vivo*. Cell 111:41–50

Funk D, Amir S (1999) Conditioned fear attenuates light-induced suppression of melatonin release in rats. Physiol Behav 67:623–626

Gall D, Bieske K (2004) Definition and measurement of circadian radiometric quantities. Proceedings of the CIE Symposium '04 on Light and Health Vienna: Commission Internationale de l'Éclairage 2004, pp 129–132

Garcia-Maurino S, Pozo D, Carrillo-Vico A, Calvo JR, Guerrero JM (1999) Melatonin ac-tivates Th1 lymphyocytes by increasing IL-12 production. Life Sci 2143–2150

Gauger MA, Sancar A (2005) Cryptochrome, circadian cycle, cell cycle checkpoints, and cancer. Cancer Res 65:6828–6834

Girardin JL, Kripke DF, Cole RJ, Elliott JA (2000) No melatonin suppression by illumination of popliteal fossae or eyelids. J Biol Rhythms 15:265–269

Gooley JJ, Lu J, Chou TC, Scammell TE, Saper CB (2001) Melanopsin in cell of origin of the retinohypothalamic tract. Nat Neurosci 4:1165

Grin W, Grünberger W (1998) A significant correlation between melatonin deficiency and endometrial cancer. Gynecol Obst Invest 45:62–65

Gronfier C, Czeisler CA (2004) Efficacy of a single sequence of intermittent bright light pulses for delaying circadian phase in humans. Am J Physiol Endocrinol Metab 287:E174–E181

Gunnarsdottir H, Rafnsson V (1995) Cancer incidence among Icelandic nurses. J Occup Environ Med 37:307–312

Gurwitz D (1998) Flight attendants, breast cancer, and melatonin. Lancet 352:1389–1390

Guth SL, Massof RW, Benzschawel T (1980) Vector model for normal and dichromatic color vision. J Opt Soc Am 70:197–212

Haimov I, Arendt J (1999) Prevention and treatment of jet lag. Sleep Med Rev 3:229–240

Hamilton T (1969) Influence of environmental light and melatonin upon mammalian tu-mor induction. Br J Surg 56:764–766

Hanahan D, Weinberg RA (2000) The hallmarks of cancer. Cell 100:57–70

Hardeland R, Backhaus C, Fadavi A, Hess M (2007a) N1-acetyl-5-methoxykynuramine contrasts with other tryptophan metabolites by a peculiar type of NO scavenging: cydization to a cinnolinone prevents formation of unstable nitrosamines. J Pineal Res 43:104–105

Hardeland R, Backhaus C, Fadavi A (2007b) Reactions of the NO redbox forms of NO^+, and HNO protonated NO^- with the melatonin metabolite N1-acetyl-5-methoxykynuramine. J Pineal Res 43:382–388

Harder B (2006) Bright lights, big cancer. Science News Online. 169:8. Epub: 7 Jan 2006

Hashimoto S, Nakamura K, Honma S et al (1996) Melatonin rhythm is not shifted by light that suppress nocturnal melatonin in humans under entrainment. Am J Physiol 270:R1073–R1077

Hattar S, Liao H-W, Takao M, Berson DM, Yau K-W (2002) Melanopsin-containing retinal ganglion cells: architecture, projections, and intrinsic photosensitivity. Science 295:1065–1070

Haus E (2009) Chronobiology in oncology. Int J Radiat Oncol Biol Phys 73:3–5

Hebert M, Martin SK, Lee C, Eastman CL (2002) The effects of prior light history on the suppression of melatonin by light in humans. J Pineal Res 33:198–203

Helmholtz H (1896) Handbuch der Physiologischen Optik Hamburg: Voss

Hilakivi-Clarke L, de Assis S (2006) Fetal origins of breast cancer. TRENDS Endocrinol Metab 17:340–348

Hill AB (1965b) The environment and disease: association or causation? Proc Royal Soc Med 58:295–300

Horne JAO, Ostberg O (1976) A self-assessment questionnaire to determine waxler morningness-eveningness in human circadian rhythms. Int J Chronobiol 4:97–110

Jasser SA, Blask DE, Brainard GC (2006a) Light during darkness and cancer: relationship in circadian photoreception and tumor biology. Cancer Causes Control 17:515–523

Jasser SA, Hanifin JP, Rollag MD, Brainard GC (2006b) Dim light adaptation attenuates acute melatonin suppression in humans. J Biol Rhythms 21:394–404

Jewett ME, Rimmer DW, Duffy JF, Klerman EB, Kronauer RE, Czeisler CA (1997) Human circadian pacemaker is sensitive to light throughout subjective day without evidence of transients. Am J Physiol 273:R1800–R1809

Jung B, Ahmad N (2006) Melatonin in cancer management: progress and promise. Cancer Res 66:9789–9793

Karasek M, Kowalski AJ, Suzin J, Zylinska K, Swietoslawski J (2005) Serum melatonin circadian profiles in women suffering from cervical cancer. J Pineal Res 39:73–76

Karbownik M, Reiter RJ, Cabrera J, Garcia JJ (2001) Comparison of the protective effect of melatonin with other antioxidants in a hamster kidney model of estradical-induced DNA damage. Mutat Res 474:87–92

Karbownik M, Reiter RJ, Qi W, Garcia JJ, Tan DX, Manchester LC, Vijayalaxmi (2000) Protective effects of melatonin against oxidation of quanine bases in DNA and decreased microsomal fluidity in rat liver induced by whole body ionizing radiation. Mol Cell Biochem 211:137–144

Kennaway DJ, Moyer RW, Voultsios A et al (2001) Serotonin, excitatory amino acids and the photic control of melatonin rhythms and SCN c-FOS in the rat. Brain Res 897:36–43

Kennaway DJ, Rowe SA (2000) Effect of stimulation of endogenous melatonin secretion during constant light exposure on 6-sulphatoxymelatonin rhythmicity in rats. J Pineal Res 28:16–25

Kerenyi NA, Pandula E, Feuer G (1990) Why the incidence of cancer is increasingly: the role of "light pollution". Med Hypotheses 33:75–78

Kesteloot H (2004) Alcohol intake and markers of inflammation. Eur Heart J 25:2075–2076

Kiefer TL, Lai L, Yuan L, Dong C, Burow ME, Hill SM (2005) Differential regulation of estrogen receptor alpha, glucocortoid receptor and retinoic acid receptor alpha transcriptional activity by melatonin is mediated via different G proteins. J Pineal Res 38:231–239

Kliukiene J, Tynes T, Andersen A (2001) Risk of breast cancer among Norwegian women with visual impairment. Br J Cancer 84:397–399

Knutson KL, Spiegel K, Penev P, Van Cauter E (2007) The metabolic consequences of sleep deprivation. Sleep Med Rev 11:163–178

Ko CH, Takahashi JS (2006) Molecular components of the mammalian circadian clock. Hum Mol Genet 15:R271–R277

Kreier F, Kalsbeek A, Sauerwein HP, Fliers E, Romijn JA, Buijs RM (2007) "Diabetes of the elderly" and type 2 diabetes in younger patients: possible role of the biological clock. Exp Gerontol 42:22–27

Kubo T, Ozasa K, Mikami K (2006b) Prospective cohort study of the risk of prostate cancer among rotating-shift workers: findings from the Japan collaborative cohort study. Am J Epidemiol 164:549–555

Kubota T, Uchiyama M, Suzuki H et al (2002) Effects of nocturnal bright light on saliva melatonin, core body temperature and sleep propensity rhythms in human subjects. Neurosci Res 42:115–122

Laakso M-L, Hatonen T, Stenber D et al (1993) One-hour exposure to moderate illuminance (500 lux) shifts the human melatonin rhythm. J Pineal Res 15:21–26

Laposky AD, Bass J, Kohsaka A, Turek FW (2008) Sleep and circadian rhythms: key components in the regulation of energy metabolism. FEBS Lett 582:142–151

Leon-Blanco MM, Guerrero JM, Reiter RJ, Calvo JR, Pozo D (2003) Melatonin inhibits telomerase activity in the MCF-7 tumor cell line both *in vivo* and *in vitro*. J Pineal Res 35:204–211

Lewy AJ, Sack RL, Miller S et al (1987) Antidepressant and circadian phase-shifting effects of light. Science 235:352–354
Lissoni P, Bolis S, Brivio F, Fumagalli L (2000) A phase II study of neuroimmunotherapy with subcutaneous low-dose IL-2 plus pineal hormone melatonin in untreatable advanced hematologic malignancies. Anticancer Res 20:2103–2105
Lissoni P (2002) Is there a role for melatonin in supportive care? Support Cancer Care 10:110–116
Lockley SW, Brainard GC, Czeisler CA (2003) High sensitivity of the human circadian melatonin rhythm to resetting by short wavelength light. J Clin Endocrinol Metab 88:4502–4505
Lockley SW, Evans EE, Scheer FA, Brainard GC, Czeisler CA, Aeschbach D (2006) Short-wavelength sensitivity for the direct effects of light on alertness, vigilance, and the waking electroencephalogram in humans. Sleep 29:161–168
Lockley SW, Skene DJ, Arendt J, Tabandeh H, Bird AC, Defrance R (1997) Relationship between melatonin rhythms and visual loss in the blind. J Clin Endocrinol Metab 82:3767–3770
Lucas RJ (2006) Chromophore regeneration: melanopsin does its own thing. Proc Natl Acad Sci USA 103:10153–10154
Lynge E, Thygesen L (1990) Occupational cancer in Denmark: cancer incidence in the 1970 census population. Scand J Work Environ Health 16(Suppl 2):1–35
Lynge E (1996) Risk of breast cancer is also increased among Danish female airline cabin attendants. Br Med J 312:253
Ma X, Idle JR, Krauz KW, Tan DX, Ceraulo L, Gonzalez FJ (2006) Urinary metabolites and antioxidant products of exogenous melatonin in the mouse. J Pineal Res 40:343–349
Maemura K, Takeda N, Nagai R (2007) Circadian rhythms in the CNS and peripheral clock disorders: role of the biological clock in cardiovascular diseases. J Pharmacol Sci 103:134–138
Maestroni GJ (1993) The immunoneuroendocrine role of melatonin. J Pineal Res 14:1–10
Malins DC, Polissar NL, Gunselman SJ (1996) Progression of human breast cancers to the metastatic state is linked to hydroxyl radical-induced DNA damage. Proc Natl Acad Sci USA 93:2557–2563
Manda K, Ueno M, Anzai K (2007) AFMK, a melatonin metabolite, attenuates X-ray-induced oxidative damage to DNA, proteins and lipids in mice. J Pineal Res 42:386–393
Martinez-Campa, Alonzo-Gonzalez C, Mediavilla MD, Cos S, Gonzalez A, Ramos S, Sanchez-Barcelo EJ (2006) Melatonin inhibits both ERα activation and breast cancer cell proliferation induced by the metalloestrogen, cadmium. J Pineal Res 40:291–296
Mayo JC, Sainz RM, Tan DX, Antolin I, Rodriguez C. Reiter RJ (2005) Melatonin and Parkinson's disease. Endocrine 27:161–178
McIntyre IM, Norman TR, Burrows GD, Armstrong SM (1989b) Quantal melatonin suppression by exposure to low intensity light in man. Life Sci 45:327–332
Mediavilla MD, Cos S, Sanchez-Barcelo EJ (1999) Melatonin increases p53 and p21WAF1 expression in MCF-7 human breast cancer cells *in vitro*. Life Sci 65:415–420
Mendoza J (2007) Circadian clocks: setting time by food. J Neuroendocrinol 19:127–137
Miller AB, Gaudette LA (1996) Breast cancer in circumpolar Inuit, 1969–1988. Acta Oncol 35:577–580
Miller D, Bierman A, Figueiro MG, Schernhammer ES, Rea MS (2009) Ecological measurements of light exposure, activity, and circadian disruption in real-world environments. Proceedings of experiencing light 2009. Eindhoven University of Technology, Eindhoven, 2009, pp 53–61
Mills E, Wu P, Seely D, Guyatt G (2005) Melatonin in the treatment of cancer: a systematic review of randomized controlled trials and meta-analysis. J Pineal Res 39:360–366
Moore RY, Leak RI (2001) Suprachiasmatic nucleus. In: Takahashi JS, Turek FW, Moore RY (eds) Handbook of behavioral neurobiology. vol 12: Circadian Clocks. Kluwer Academic, New York , pp 141–171

National Sleep Foundation (2005) The 2005 NSF National Sleep in America Poll. Available. http://www.sleepfoundation.org
Panda S, Provencio I, Tu DC, Pires SS, Rollag MD, Castrucci AM et al (2003) Melanopsin is required for non-image-forming photic responses in blind mice. Science 301:525–527
Pinheiro SP, Schernhammer ES, Tworoger SS, Michels KB (2006) A prospective study of habitual duration of sleep and incidence of breast cancer in a large cohort of women. Cancer Res 66:5521–5525
Plautz JD, Kaneko M, Hall JC, Kay SA (1997) Independent photoreceptive circadian clocks throughout Drosophila. Science 278:1632–1635
Poeggeler B (2005) Melatonin, aging and age-related diseases. Endocrine 27:201–212
Poole C (2002) The darkness at the end of the tunnel: summary and evaluation of the international symposium on light endocrine systems and cancer. Neuroendocrinol Lett 23(Suppl2):71–78
Prener A, Storm HH, Nieken NH (1996) Cancer of the male genital tract in circumpolar Inuit. Acta Insol 35:589–593
Provencio I, Jiang G, De Grip WJ, Hayes WP, Rollag MD (1998) Melanopsin: an opsin in melanophores, brain, and eye. Proc Natl Acad Sci USA 95:340–345
Provencio I, Rodriguez IR, Jiang G, Hayes WP, Moreira EF, Rollag MD (2000) A novel human opsin in the inner retina. J Neurosci 20:600–605
Provencio I, Rollag MD, Castrucci AM (2002) Photoreceptive net in the mammalian retina. Nature 415:493
Pu M (2000) Physiological response properties of cat retinal ganglion cells projecting to suprachiasmatic nucleus. J Biol Rhythms 15:31–36
Pukkala E, Ojamo M, Rudanko SL, Stevens RG, Verkasalo PK (2006) Does incidence of breast cancer, and prostate cancer decrease with increasing degree of visual impairment. Cancer Causes Control 17:573–576
Pukkala E, Verkasalo PK, Ojarmo M, Rudanko SL (1999) Visual impairment and cancer: a population-based cohort study in Finland. Cancer Causes Control 10:13–20
Qi W, Reiter RJ, Tan DX, Garcia JJ, Manchester LC, Karbownik M, Calvo JR (2000a) Chromium (III) induced 8-hydroxydeoxyguanosine in DNA and its reduction by antioxidants: Comparative effects of melatonin, ascorbate and vitamin E. Environ Health Persp 108:2399–2402
Qi W, Reiter RJ, Tan DX, Manchester LC, Siu AW, Garcia JJ (2000b) Increased levels of oxidatively damaged DNA induced by chromium (III) and H_2O: protection by melatonin and related molecules. J Pineal Res 29:54–61
Rajaratnam SMW, Arendt J (2001) Health in a 24-h society. Lancet 358:999–1005
Ralph MR, Foster RG, Davis FC, Menaker M (1990) Transplanted suprachiasmatic nucleus determines circadian period. Science 247:975–978
Rea MS, Bierman A, Figueiro MG, Bullough JD (2008) A new approach to understanding the impact of circadian disruption on human health. J Circadian Rhythms 6:7–20
Rea MS, Bullough JD, Figueiro MG (2001) Human melatonin suppression by light: a case for scotopic efficiency. Neurosci Lett 299:45–48
Rea MS, Bullough JD, Figueiro MG (2002) Phototransduction for human melatonin suppression. J Pineal Res 32:209–213
Rea MS, Bullough JD, Freyssinier-Nova JP, Bierman A (2004) A proposed unified system of photometry. Lighting Res Technol 36:85–111
Rea MS (ed) (2000) IESNA lighting handbook: reference and application. Illuminating Engineering Society of North America, New York
Rea MS, Figueiro MG, Bierman A, Bullough JD (2010) Circadian light. J Circadian Rhythms 8:2
Rea MS, Figueiro MG, Bullough JD, Bierman A (2005) A model of phototransduction by the human circadian system. Brain Res Rev 50:213–228

Reiter RJ, Tan DX, Korkmaz A, Erren TC, Piekarski C, Tamura H, Manchester LC (2007b) Light at night chronodisruption, melatonin suppression, and cancer risk: a review. Critical Rev Oncogenesis 13(4):303–328
Reiter RJ, Tan DX, Sainz RM, Mayo JC (2002a) Melatonin protects the heart against both ischemia/reperfusion injury and chemotherapeutic agents. Cardiovasc Drug Ther 16:5–6
Reiter RJ, Tan DX, Sainz RM, Mayo JC (2002b) Melatonin: reducing the toxicity and in-creasing the efficacy of drugs. J Pharm Pharmacol 54:1299–1321
Reiter RJ (2004) Mechanisms of cancer inhibition by melatonin. J Pineal Res 37:213–214
Reiter RJ (2000) Melatonin: lowering the high price of free radicals. News Physiol Sci 15:246–250
Reiter RJ (1991) Melatonin: the chemical expression of darkness. Mol Cell Endocrinol 79:C153–C158
Reiter RJ (1993) The mammalian pineal gland as an end organ of the visual system. In: Wetterberg L (ed). Light and biological rhythms in man. Elsevier, Amsterdam, pp 145–160
Reiter RJ (1995) The pineal gland and melatonin in relation to aging: a summary of the theories and of the data. Exp Gerontol 30:119–212
Rennert G, Pinchev M, Rennert HS (2010) Use of bisphosphonates and risk of postmenopausal breast cancer. J Clin Oncol 28:3577–3581
Reuss S (2003) The clock in the brain: anatomy of the mammalian circadian timing system. Endokrinologie 6:9–48
Revell VL, Arendt J, Terman M, Skene DJ (2005) Short-wavelength sensitivity of the human circadian system to phase-advancing light. J Biol Rhythms 20:270–272
Revell VL, Skene DJ (2007) Light-induced melatonin suppression in humans with polychromatic and monochromatic light. Chronobiol Int 24:1125–1137
Rix BA, Lange E (1996) Cancer incidence in Danish health care workers. Scand J Soc Med 24:114–120
Rodriguez C, Mayo JC, Sainz RM, Antolin I, Herrera F, Martin V, Reiter RJ (2004) Regulation of antioxidant enzymes: a significant role of melatonin. J Pineal Res 36:1–9
Rosen J, Than NN, Koch D, Poeggeler B, Laatsch H. Hardeland R (2006) Interactions of melatonin and its metabolites with the ABTS cation radical: extension of the radical scavenging cascade and formation of a novel class of oxidation products, C2-substituted indolinones. J Pineal Res 41:374–381
Saini VD, Cohen GH (1979) Using color substitution pupil response to expose chromatic mechanisms. J Opt Soc Am 69:1029–1035
Sainz RM, Mayo JC, Rodriguez C, Tan DX, Lopez-Burrillo S, Reiter RJ (2003) Melatonin and cell death: differential actions on apoptosis in normal and cancer cells. Cell Mol Life Sci 60:407–426
Sainz RM, Mayo JC, Tan DX, Leon J, Manchester LC, Reiter RJ (2005) Melatonin reduces prostate cancer cell growth leading to neuroendocrine differentiation via a receptor and PKA independent mechanism. Prostate 63:29–43
Sanchez-Barcelo E, Cos S, Mediavilla D, Martinez-Campa C, Gonzalez A, Alonso-Gonzalez C (2005) Melatonin-estrogen interactions in breast cancer. J Pineal Res 38:217–222
Sanchez-Barcelo EJ, Cos S, Fernandez R, Mediavilla MD (2003) Melatonin and mammary cancer: a short review. Endocrine Related Cancer 10:153–159
Sankila R, Karjalainen S, Laara E, Pukkala E, Oksanen H, Hakulain T, Teppo L, Hakama M (1990) Cancer risk among health care personnel in Finland, 1971–1980. Scand J Work Environ Health 16:252–257
Sauer LA, Dauchy RT, Blask DE (2001) Polyunsaturated fatty acids, melatonin and cancer prevention. Biochem Pharmacol 61:1455–1462
Schernhammer ES, Hankinson SE (2005) Urinary melatonin levels and breast cancer risk. J Natl Cancer Inst 97:1084–1087
Schernhammer ES, Kroenke CH, Laden F, Hankinson SE (2006) Night work and risk of breast cancer. Epidemiology 17:108–111

Schmidt TM, Kofuji P (2009) Functional and morphological differences among intrinsically photosensitive retinal ganglion cells. J Neurosci 29:476–482

Sedjo RL, Byers T, Barrera E Jr, Cohen C, Fontham ET, Newman LA, Runowicz CD, Thorson AG, Thun MJ et al () A midpoint assessment of the American Cancer

Sephton S, Spiegel D (2003) Circadian disruption in cancer: a neuroendocrine-immune pathway from stress to disease? Brain Behav Immun 17:321–328

Shiu SYW (2007) Towards rational and evidence-based use of melatonin in prostate can-cer prevention and treatment. J Pineal Res 43:1–9

Skene D, Bojkowski CJ, Currie JE, Wright J (1990) 6-sulfatoxymelatonin production in breast cancer patients. J Pineal Res 8:269–276

Smith VC, Pokorny J (1975) Spectral sensitivity of the foveal cone photopigments between 400 and 500 nm. Vision Res 15:161–171

Snyder OB, Kelly JJ, Lanier AP (2006) Prostate cancer in Alaskan natives, 1969–2003. Int J Circumpol Health 65:8–17

Spiegel K, Knutson K, Leproult R, Tasali E, Van Cauter E (2005) Sleep loss: a novel risk factor for insulin resistance and Type 2 diabetes. J Appl Physiol 99:2008–2019

Spiegel K, Tasali E, Penev P, Van Cauter E (2004) Sleep curtailment in healthy young men is associated with decreased leptin levels, elevated ghrelin levels, and increased hunger and appetite. Ann Intern Med 141:846–850

Srinivasan V, Smits M, Spence W, Lowe AD, Kayumov L, Pandi- Perumal SR et al (2006) Melatonin in mood disorders. World J Biol Psychiatry 7:138–151

Srinivasan V, Spence D, Pandi-Perumal S, Trakht I, Esquifino A, Cardinali D, Maestroni G (2008) Melatonin, environmental light, and breast cancer. Breast Cancer Res Treat 108:339–350

Stevens RG (1987a) Electric power use and breast cancer: a hypothesis. Am J Epidemiol 125:556–561

Stevens RG (2005b) Circadian disruption and breast cancer: from melatonin to clock genes. Epidemiology 16:254–258

Stevens RG (2006) Artifcial lighting in the industrialized world: circadian disruption and breast cancer. Cancer Causes Control 17:501–507

Stevens RG (1987b) Electric power use and breast cancer: a hypothesis. Am J Epidemiol 125:556–561

Stevens RG (2002) Lighting during the day and night: possible impact on risk of breast cancer. Neuroendocrinol Lett 23(Suppl 2):57–60

Stokkan KA, Reiter RJ (1994) Melatonin rhythms in Arctic urban residents. J Pineal Res 16:33–36

Storm HH, Nielsen NH, Prener A, Jensen OM (1991) A comparison of cancer in Greenland and Denmark: a study based on routinely collected incidence data 1973–1985, using Danish population as a baseline. Circumpolar Health 90:470–471

Takahashi JS, De Coursey PJ, Bauman L et al (1984) Spectral sensitivity of a novel photoreceptive system mediating entrainment of mammalian circadian rhythms. Nature 308:186–188

Tan DX, Manchester LC, Terron MP, Flores LJ, Reiter RJ (2007) One molecule, many derivatives: a never-ending interaction of melatonin with reactive oxygen and nitrogen species. J Pineal Res 42:28–42

Tan DX, Poeggeler B, Reiter RJ, Chen LD, Chen S, Manchester LC, Barlow-Walden LR (1993) The pineal hormone melatonin inhibits DNA-addict formation induced by the chemical carcinogen safrole *in vivo*. Cancer Lett 70:65–71

Tan DX, Reiter RJ, Chen LD, Poeggeler B, Manchester LC, Barlow-Walden LR (1994) Both physiological levels of melatonin reduces DNA addict formation induced by the carcinogen safrole. Carcinogenesis 15:215–218

Tengattini S, Reiter RJ, Tan DX, Terron MP, Rodella LF, Rezzani R (2008) Cardiovascular disease: protective effects of melatonin. J Pineal Res 44:16–25

Thapan K, Arendt J, Skene DJ (2001) An action spectrum for melatonin suppression: evidence for a novel non-rod, noncone photoreceptor system in humans. J Physiol 535:261–267
Tosini G, Menaker M (1996) Circadian rhythms in cultured mammalian retina. Science 272:419–421
Van Gool SW, VanDen HL, Ceuppens JL (2000) Activation of the immune system in cancer patients. Med Pediatr Oncol 34:1–9
Verkasalo PK, Lillberg K, Steven RG, Hublin C, Patinen M, Koskenvuo M, Kaprio J (2005) Sleep duration and breast cancer: a prospective cohort study. Cancer Res 65:9595–9600
Vijayalaxmi, Thomas CR, Reiter RJ, Herman TS (2002) Melatonin: from basic science to cancer treatment clinics. J Clin Oncol 20:2575–2601
Viswanathan AN, Hankinson SE, Schernhammer ES (2007) Night shift work and the risk of endometrial cancer. Cancer Res 67:10618–10622
Voultsios A, Kennaway DJ, Dawson D (1997) Salivary melatonin as circadian phase marker: validation and comparison to plasma melatonin. J Biol Rhythms 12:457–465
Warman V, Difk D-J, Warman GR, Arendt J, Skene D (2003) Phase advancing human circadian rhythms with short wavelength light. Neurosci Lett 342:37–40
Wartenberg D, Stapleton CP (1998) Risk of breast cancer is also increased among US female airline cabin attendants. Br Med J 316:1902
Waxler M, James RH, Brainard GC et al (1992) Retinopathy and bright light therapy. Am J Psychiatry 149:1610–1611
Weale RA (1953) Spectral sensitivity and wave-length discrimination of the peripheral retina. J Physiol 119:170–190
Wehr TA (1992b) In short days human sleep is biphasic. Sleep Res 1:103–107
Wehr TA (2001) Photoperiodism in humans and other primates: evidence and implications. J Biol Rhythms 16:348–364
Wehr TA (1991) The durations of human melatonin secretion and sleep respond to changes in day length (photoperiod). J Clin Endocrinol Metab 73:1276–1280
Whelan EM, Ross GL, Stimola AN (eds) American's war on "carcinogens". American Council on Science and Health, New York, p 177
Witt-Enderby PA, Radio NM, Doctor JS, Davis VL (2006) Therapeutic treatments potentially medicated by melatonin receptors: potential clinical uses in the prevention of osteoporosis, cancer and as a adjuvant therapy. J Pineal Res 41:297–305
Wright HR, Lack LC (2001) Effect of light wavelength on suppression and phase delay of the melatonin rhythm. Chronobiol Int 18:801–808
Yamazaki S, Numano R, Abe M, Hida A, Takahashi R, Ueda M et al (2000) Resetting central and peripheral circadian oscillators in transgenic rats. Science 288:682–685
Young ME, Bray MS (2007) Potential role for peripheral circadian clock dyssynchrony in the pathogenesis of cardiovascular dysfunction. Sleep Med 8:656–667
Zeitzer JM, Kronauer RE, Czeisler CA (1997) Photopic transduction implicated in human circadian entrainment. Neurosci Lett 232(135–138):19
Zhu Y, Brown HN, Zhang Y, Stevens RG, Zheng T (2005) Period structural variation: a circadian biomarker associated with breast cancer in young women. Cancer Epidemiol Biomarker Prev 14:268–270

Index

A. Haim and B. A. Portnov, *Light Pollution as a New Risk Factor for Human Breast and Prostate Cancers*, DOI: 10.1007/978-94-007-6220-6,

MIX
Papier aus verantwortungsvollen Quellen
Paper from responsible sources
FSC® C105338

If you have any concerns about our products,
you can contact us on
ProductSafety@springernature.com

In case Publisher is established outside the EU,
the EU authorized representative is:
Springer Nature Customer Service Center GmbH
Europaplatz 3, 69115 Heidelberg, Germany

Printed by Libri Plureos GmbH
in Hamburg, Germany